Antimicrobial Resistance in Horses

Antimicrobial Resistance in Horses

Editors

Amir Steinman
Shiri Navon-Venezia

MDPI • Basel • Beijing • Wuhan • Barcelona • Belgrade • Manchester • Tokyo • Cluj • Tianjin

Editors
Amir Steinman
The Hebrew University of Jerusalem
Israel

Shiri Navon-Venezia
Ariel University
Israel

Editorial Office
MDPI
St. Alban-Anlage 66
4052 Basel, Switzerland

This is a reprint of articles from the Special Issue published online in the open access journal *Animals* (ISSN 2076-2615) (available at: https://www.mdpi.com/journal/animals/special_issues/ antimicrobial_resistance_horses).

For citation purposes, cite each article independently as indicated on the article page online and as indicated below:

LastName, A.A.; LastName, B.B.; LastName, C.C. Article Title. *Journal Name* **Year**, *Article Number*, Page Range.

ISBN 978-3-03936-712-2 (Hbk)
ISBN 978-3-03936-713-9 (PDF)

Cover image courtesy of Dalia Berlin.

Contents

About the Editors

Amir Steinman (Ph.D.) holds a Degree in Veterinary Medicine (DVM, 1996), a PhD (2008) on the immune response of cattle to botulism, and a master's degree (2010) in health administration. Dr. Steinman served as Head of the large animal department (2006–2013), veterinary teaching hospital, Koret School of Veterinary Medicine (KSVM-VTH). Since 2013, he has served as Director of KSVM-VTH. His research is focused on equine infectious diseases, mainly vector-borne diseases and antimicrobial resistance.

Shiri Navon-Venezia (Ph.D.) is a Full Prof. in Microbiology, Head of the Laboratory of Bacterial Pathogens and Antibiotic Resistance at the Department of Molecular Biology and at the Sheldon School of Medicine, Ariel University. Prof. Navon-Venezia is studying bacterial pathogens of clinical importance to decipher the molecular mechanisms that lead to antibiotic resistance, resistance spread, and bacterial virulence in order to develop improved diagnostics and new therapies against multidrug resistant.

Editorial

Antimicrobial Resistance in Horses

Amir Steinman [1,*] **and Shiri Navon-Venezia** [2,*]

[1] Koret School of Veterinary Medicine, The Hebrew University of Jerusalem, Rehovot 7610001, Israel
[2] Department of Molecular Biology and the Adelson School of Medicine, Ariel University, Ariel 4077625, Israel
* Correspondence: amirst@savion.huji.ac.il (A.S.); shirinv@ariel.ac.il (S.N.-V.)

Received: 22 June 2020; Accepted: 25 June 2020; Published: 9 July 2020

Antimicrobial resistance (AMR) is an increasingly recognized global public health threat to the modern health-care system that could hamper the control and treatment of infectious diseases [1]. Microorganisms may serve as a reservoir for AMR in all ecological niches; therefore, a "one health" coordinated multisectorial approach is desired to investigate and address this warning phenomenon [2]. This approach appears to be a winning strategy to combat and reduce the burden of AMR, but it requires combined forces and resources that are consistently and effectively implemented by both human and veterinary health professionals [1].

Horses are among the most central animals in human history; they have been used in wars, as a means of transport, and even facilitated work in mines. Since then, the rate of contact between domesticated horses and humans has steadily increased. Nowadays, horses play an important role as sport animals and in animal-assisted therapy. Due to these close human-horse interactions, the adequate detection of infectious diseases and AMR that may affect both humans and horses is crucial, especially in cases of highly transmissible diseases [3]. Numerous important antibiotic-resistant zoonotic pathogens have been reported from horses, including extended-spectrum beta-lactamases (ESBL)-producing *Escherichia coli*, methicillin-resistant *Staphylococcus aureus* (MRSA), and multidrug-resistant (MDR) *Salmonella*. These reports have attracted increasing attention to the threat of AMR in horses [4].

During the last two decades, researchers have generated a vast amount of information on the importance of MRSA in horses, which has been recognized as an occupational risk for veterinary professionals [5]. MRSA outbreaks affecting both horses and personnel were reported from different geographic locations and reciprocal animal-personnel transmission of infections was demonstrated. Furthermore, it was previously demonstrated that on-admission MRSA colonization in horses is a risk factor to develop MRSA infection [6]. In spite of the accumulating data on the prevalence, risk factors for colonization, and resistance genes of ESBL-producing *Enterobacteriaceae*, data that links between resistant gram-negative gut colonization and equine health is still lacking.

The occurrence of AMR pathogens causing infections in equine populations increases concern over the issue of antimicrobial stewardship that involves the judicious use of antimicrobials balanced with the requirement to treat the presenting clinical condition [7]. The challenges in equine practice include the size and value of the patient, correct and timely pathogen identification, and its susceptibility profile, together with the limited number of drugs and their indiscriminate use by clients [7]. Therefore, it is crucial to promote antimicrobial stewardship, not just among academics, public health personnel, and specialists, but also among primary care equine clinicians and equine caretakers [8].

Another important aspect of AMR in horses is the proper use of critically important antibiotics (CIA) such as fluoroquinolones, third and fourth generation cephalosporins, and macrolides. The prophylactic use of macrolide with rifampin in foals suspected to be infected with *Rhodococcus equi* has been shown to promote MDR in both *R. equi* and in gut commensals, increasing the risk of environmental shedding [9]. Disease-specific practice guidelines are required to reduce CIA use for skin, respiratory, and postsurgical infections in equine medicine [10]. Therefore, as equine practitioners and researchers, we should pay attention to the use of CIAs in equine patients treatment [1].

The aim of this special issue on AMR in horses was to collect the most recent data on the prevalence, risk factors, and characterization of different MDR pathogens in different equine cohorts from various countries. Data from Israel reports on colonization with ESBL-producing *Enterobacteriaceae* in foals on admission and in the hospital setting. ESBL colonization in neonatal foals was associated with umbilical infection and ampicillin treatment during hospitalization [11]. In Israel, risk factors for ESBL-E shedding in farm horses included horses' breed, sex, and previous antibiotic treatment [12]. In a similar cohort of healthy horses from Canada, the number of staff members and equestrian event participation were identified as risk factors for MDR *E. coli* shedding [13]. In a study from Japan, healthy racehorses were reported to be colonized with MDR ESBL/AmpC-producing *Klebsiella pneumoniae* [14]. Another unique horse population that AMR pathogens were recovered from was equine destined for human consumption in Spain, in which both nasal and fecal carriage of a highly virulent MRSA was detected [15].

In addition, ESBL-producing *Enterobacteriacae* pathogens were also reported as causative agents of clinical infections in horses. In France, the percentages of MDR *Staphylococcus aureus* and MDR *Enterobacter* spp. strains causing clinical infections increased significantly during a 3-year period [16]. In Austria, MDR *Klebsiella* species were isolated from clinical samples, displaying a variety of resistance and virulence genes [17]. In a clinical bacterial collection from Texas-A&M, ESBL-producing *Enterobacteriacae* were reported with the first report of *E. coli* ST1308 in horses [18]. We believe that the new data reported here is highly relevant from a "one health" perspective; it will help to improve our knowledge related to the issue of AMR worldwide and will assist in improving control measures, optimize appropriate therapy, and will encourage further studies in this important field.

Funding: This research received no external funding.

Conflicts of Interest: The authors declare no conflict of interest.

References

1. Ferri, M.; Ranucci, E.; Romagnoli, P.; Giaccone, V. Antimicrobial resistance: A global emerging threat to public health systems. *Crit. Rev. Food Sci. Nutr.* **2017**, *57*, 2857–2867. [CrossRef] [PubMed]
2. Palma, E.; Tilocca, B.; Roncada, P. Antimicrobial Resistance in Veterinary Medicine: An Overview. *Int. J. Mol. Sci.* **2020**, *21*, 1914. [CrossRef] [PubMed]
3. Lönker, N.S.; Fechner, K.; Wahed, A.A.E. Horses as a crucial part of one health. *Vet. Sci.* **2020**, *7*, 28. [CrossRef]
4. Isgren, I. Antimicrobial resistance in horses. *Vet. Rec.* **2018**, *183*, 316–318.
5. Hanselman, B.A.; Kruth, S.A.; Rousseau, J.; Low, D.E.; Willey, B.M.; McGeer, A.; Weese, J.S. Methicillin-resistant Staphylococcus aureus colonization in veterinary personnel. *Emerg. Infect. Dis.* **2006**, *12*, 1933–1938. [CrossRef]
6. Weese, J.S.; Rousseau, J.; Willey, B.M.; Archambault, M.; McGeer, A.; Low, D.E. Methicillin-resistant Staphylococcus aureus in horses at a veterinary teaching hospital: Frequency, characterization, and association with clinical disease. *J. Vet. Intern. Med.* **2006**, *20*, 182–186. [CrossRef] [PubMed]
7. Raidal, S.L. Antimicrobial stewardship in equine practice. *Aust. Vet. J.* **2019**, *97*, 238–242. [CrossRef] [PubMed]
8. Weese, J.S. Antimicrobial use and antimicrobial resistance in horses. *Equine Vet. J.* **2015**, *47*, 747–749. [CrossRef] [PubMed]
9. Álvarez-Narváez, S.; Berghaus, L.J.; Morris, E.R.A.; Willingham-Lane, J.M.; Slovis, N.M.; Giguere, S.; Cohen, N.D. A Common Practice of Widespread Antimicrobial Use in Horse Production Promotes Multi-Drug Resistance. *Sci. Rep.* **2020**, *10*, 911. [CrossRef] [PubMed]
10. Lhermie, G.; La Ragione, R.M.; Weese, J.S.; Olsen, J.E.; Christensen, J.P.; Guardabassi, L. Indications for the use of highest priority critically important antimicrobials in the veterinary sector. *J. Antimicrob. Chemother.* **2020**, *75*, 1671–1680. [CrossRef] [PubMed]

11. Shnaiderman-Torban, A.; Paitan, Y.; Arielly, H.; Kondratyeva, K.; Tirosh-Levy, S.; Abells Sutton, G.; Navon-Venezia, S.; Steinman, A. Extended-spectrum β-lactamase-producing *Enterobacteriaceae* in hospitalized neonatal foals: Prevalence, risk factors for shedding and association with infection. *Animals* **2019**, *9*, 600. [CrossRef] [PubMed]

12. Shnaiderman-Torban, A.; Navon-Venezia, S.; Dor, Z.; Paitan, Y.; Arielly, H.; Abu Ahmad, W.; Kelmer, G.; Fulde, M.; Steinman, A. Extended-spectrum β-lactamase-producing *Enterobacteriaceae* shedding in farm horses versus hospitalized horses: Prevalence and risk factors. *Animals* **2020**, *10*, 282. [CrossRef] [PubMed]

13. de Lagarde, M.; Fairbrother, L.M.; Arsenault, J. Prevalence, Risk Factors, and Characterization of Multidrug Resistant and ESBL/AmpC Producing *Escherichia coli* in Healthy Horses in Quebec, Canada, in 2015–2016. *Animals* **2020**, *10*, 523. [CrossRef] [PubMed]

14. Sukmawinata, E.; Uemura, R.; Sato, W.; Thu Htun, M.; Sueyoshi, M. Multidrug-Resistant ESBL/AmpC-Producing *Klebsiella pneumoniae* Isolated from Healthy Thoroughbred Racehorses in Japan. *Animals* **2020**, *10*, 639. [CrossRef] [PubMed]

15. Mama, O.M.; Gómez, P.; Ruiz-Ripa, L.; Gómez-Sanz, E.; Zarazaga, M.; Torres, C. Antimicrobial Resistance, Virulence, and Genetic Lineages of Staphylococci from Horses Destined for Human Consumption: High Detection of *S. aureus* Isolates of Lineage ST1640 and Those Carrying the lukPQ Gene. *Animals* **2019**, *9*, 900. [CrossRef] [PubMed]

16. Léon, A.; Castagnet, S.; Maillard, k.; Paillot, R.; Jean-Christophe Giard, J.C. Evolution of In Vitro Antimicrobial Susceptibility of Equine Clinical Isolates in France between 2016 and 2019. *Animals* **2020**, *10*, 812. [CrossRef] [PubMed]

17. Loncaric, I.; Rosel, A.C.; Szostak, M.P.; Licka, T.; Allerberger, F.; Ruppitsch, W.; Spergser, J. Broad-Spectrum Cephalosporin-Resistant *Klebsiella* spp. Isolated from Diseased Horses in Austria. *Animals* **2020**, *10*, 332. [CrossRef] [PubMed]

18. Elias, L.; Gillis, D.C.; Gurrola-Rodriguez, T.; Jeon, J.H.; Lee, J.H.; Kim, T.Y.; Lee, S.H.; Murray, S.H.; Ohta, N.; Scott, H.M.; et al. The Occurrence and Characterization of Extended-Spectrum-Beta-Lactamase-Producing *Escherichia coli* Isolated from Clinical Diagnostic Specimens of Equine Origin. *Animals* **2020**, *10*, 28. [CrossRef] [PubMed]

 *animals*

Article

Evolution of In Vitro Antimicrobial Susceptibility of Equine Clinical Isolates in France between 2016 and 2019

Albertine Léon [1,2,*], Sophie Castagnet [1], Karine Maillard [1], Romain Paillot [1,3] and Jean-Christophe Giard [2]

[1] LABÉO Frank Duncombe, 14053 CAEN, France; sophie.castagnet@laboratoire-labeo.fr (S.C.); karine.maillard@laboratoire-labeo.fr (K.M.); romain.paillot@laboratoire-labeo.fr (R.P.)
[2] Normandie Univ, UNICAEN, U2RM, 14033 Caen, France; jean-christophe.giard@unicaen.fr
[3] Normandie Univ, UNICAEN, Biotargen, 14033 Caen, France
* Correspondence: albertine.leon@laboratoire-labeo.fr; Tel.: +33-2-314-719-39

Received: 6 April 2020; Accepted: 6 May 2020; Published: 7 May 2020

Simple Summary: The emergence and the spread of antimicrobial drug resistant bacteria around the world is a major public health issue. In fact, the transmission of these bacteria from animals to humans has been already observed. In this context, the close relationships between horses and humans may contribute to cross-infection. Our objective in this study was to describe the antimicrobial susceptibility profiles of major equine pathogens over a 4-year period (2016–2019). For this purpose, more than 7800 bacterial isolates collected from horses in France with different types of infection were phenotypically analysed for their antimicrobial susceptibility. An increase in the resistance of *Staphylococcus aureus* and *Enterobacter* spp. was observed, especially between 2016 and 2019, with the percentage of multi-drug resistant strains rising from 24.5% to 37.4%, and from 26.3% to 51.7%, respectively. Our results point to the need to support veterinary antimicrobial stewardship to encourage the proper use of antibiotics.

Abstract: The present study described the evolution of antimicrobial resistance in equine pathogens isolated from 2016 to 2019. A collection of 7806 bacterial isolates were analysed for their in vitro antimicrobial susceptibility using the disk diffusion method. The most frequently isolated pathogens were group C *Streptococci* (27.0%), *Escherichia coli* (18.0%), *Staphylococcus aureus* (6.2%), *Pseudomonas aeruginosa* (3.4%), *Klebsiella pneumoniae* (2.3%) and *Enterobacter* spp. (2.1%). The majority of these pathogens were isolated from the genital tract (45.1%, n = 3522). With the implementation of two French national plans (named ECOANTIBIO 1 and 2) in 2012–2016 and 2017–2021, respectively, and a reduction in animal exposure to veterinary antibiotics, our study showed decreases in the resistance of group C *Streptococci*, *Klebsiella pneumoniae* and *Escherichia coli* against five classes, four classes and one class of antimicrobials tested, respectively. However, *Staphylococcus aureus*, *Escherichia coli* and *Enterobacter* spp. presented an increased resistance against all the tested classes, excepted for two fifths of *E. coli*. Moreover, the percentages of multi-drug resistant strains of *Staphylococcus aureus* and *Enterobacter* spp. also increased from 24.5% to 37.4% and from 26.3% to 51.7%, respectively. The data reported here are relevant to equine practitioners and will help to improve knowledge related to antimicrobial resistance in common equine pathogens.

Keywords: antibiotic resistance; horse pathogens; epidemiology

1. Introduction

Since the beginning of the 21st century, the emergence of multi-drug-resistant bacteria has become a major public health concern and a priority for all international institutions such as the World Health

Organization (WHO), the Food and Agriculture Organization of the United Nations (FAO) and the World Organization for Animal Health (or Organisation Internationale des Epizooties-OIE), which provide guidelines to mitigate the development of resistant bacteria [1–3].

In France, two governmental programmes were initiated over the 2012–2016 (ECOANTIBIO 1) and 2017–2021 (ECOANTIBIO 2) periods to reduce the veterinary use of antibiotics and to preserve the therapeutic arsenal for serious illness cases [4,5]. The objectives of the first programme were both quantitative (reduce, by 25%, the exposure of animals to antibiotics over a 5-year period) and qualitative (a reduction in the use of critical antibiotics in veterinary medicine including fluoroquinolones and last-generation cephalosporins) in order to reduce the occurrence of antimicrobial resistance, which is an international concern in terms of human and animal health [4].

The second programme focuses on incentivisation rather than regulatory measures by promoting communication, training, the use of alternatives to antibiotics, improvements in preventive measures for infectious diseases and the provision of the best tools for diagnosis and monitoring antibiotic sales and resistance [5].

In this context, several international studies have described the prevalence of resistant bacteria in equine samples in South Africa [6], Canada [7,8], Switzerland [9] and the United Kingdom [10]. Some of them reported a high level of resistance in horse's bacteria: from 26.6% to 50% of multi-drug resistant (MDR) strains [6,10], 60% and 68% of isolates were phenotypically extended-spectrum beta-lactamase (ESBL)-producing and methicillin-resistant, respectively [9].

In France, retrospective studies concerning data collected from 2006 to 2016 have been recently published and demonstrated a potential role of equids as a reservoir [11,12], and this report aims to evaluate the situation and its progression with the analysis of more than 7000 samples collected between 2016 and 2019.

2. Materials and Methods

From January 2016 to December 2019, bacterial isolates collected from horses (with suspicion of bacterial infection and prior antimicrobial treatment) by numerous practitioners in France were included in the study. Data for 2016 came from our previous manuscript [11]. Because samples came from several farms over very large areas, they could not be considered as geographically clustered. Analyses were performed in the Veterinary Microbiology diagnostics unit of the LABÉO Research and Diagnostic Institute. Strains were isolated on agar media (Columbia agar with 5% sheep blood or Columbia CNA agar with 5% sheep blood and eosin methylene blue agar). Strains were identified by Gram staining and commercially available identification systems, such as the API and VITEK 2 Compact® systems (bioMérieux, mArcy l'Etoile, France) or since 2018 (April), MALDI-TOF mass spectrometry (Microflex; Bruker Daltonics, Bremen, Germany), according to the manufacturers' instructions.

Antimicrobial susceptibility testing (AST) was performed using the disc diffusion method on Mueller–Hinton agar (enriched with 5% sheep blood for *Streptococcus* spp.) according to the recommendations of the CA-SFM/EUCAST (Comité de l'antibiogramme de la Société Française de Microbiologie/The European Committee on Antimicrobial Susceptibility Testing) [13]. The categorisations of antimicrobial susceptibility testing were carried out using the CA-SFM recommendations for antimicrobial drugs only used in veterinary medicine (as Ceftiofur, Cefquinome, Flumequine, Enrofloxacine and Marbofloxacine). For other drugs also used in human medicine, EUCAST recommendations were taken into account. After 18–24 h of incubation at 37 °C, the diameters of growth inhibition around the discs were measured using SIRSCAN (I2A, Montpellier, France) and interpreted to show bacteria were susceptible, intermediate or resistant according to CA-SFM/EUCAST clinical breakpoints. Bacteria that were categorised as "intermediate" were subsequently considered as "resistant" in our study. Bacterial isolates were evaluated for their susceptibilities to β-lactams, polymyxins, aminoglycosides, tetracycline, macrolides, rifampicin, sulphonamides and fluoroquinolones. Due to their intrinsic resistance to low levels of aminoglycosides, high concentration (HC) aminoglycosides discs were used against *Streptococci*.

Statistical analysis was performed using the XLStat software. The chi-square test was used to test for significant changes in antimicrobial resistance among each bacterial species between one year and its previous year. The temporal trends in the prevalence of antimicrobial resistance were investigated for each antimicrobial compound using the Cochran Armitage trend test. For these analyses, $p < 0.05$ were considered as significant.

3. Results

3.1. Identification and Distribution of Bacterial Isolates

In a 4-year period, 7806 bacterial isolates were included (2016: n = 1895; 2017: n =1978; 2018: n = 2125; 2019: n = 1808). These isolates were clustered from genital (45.1%; n = 3522), respiratory (22.1%; n = 1728) and cutaneous (16.3%; n = 1273) origins and other origins such as digestive or ophthalmic (16.4%; n = 1283). The most frequently isolated pathogens were group C *Streptococci* including *Streptococcus equi* subsp. *zooepidemicus*, *Streptococcus equi* subsp. *equi* and *Streptococcus dysgalactiae* subsp. *equisimilis* (27.0%, n = 2118); *Escherichia coli* (18.0%, n = 1382); *Staphylococcus aureus* (6.2%, n = 482); *Pseudomonas aeruginosa* (3.4%, n = 268); *Klebsiella pneumoniae* (2.3%, n = 180); and *Enterobacter* spp (2.1%, n = 165). The relationship between pathogen types and sampling origins (i.e., types of infection) is illustrated in Figure 1.

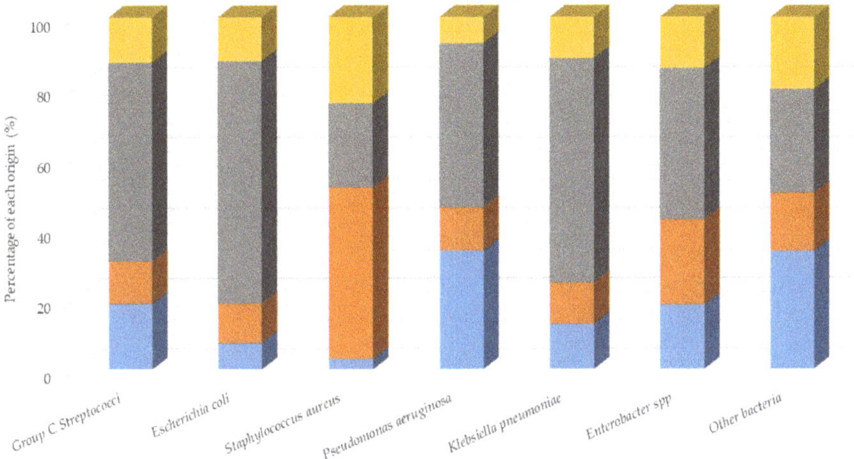

Figure 1. Repartition of sampling origins (%)—in blue, respiratory; in orange, cutaneous; in grey, genital; and in yellow, others—according to pathogen type.

3.2. Antimicrobial Susceptibility

3.2.1. GRAM Positive Bacteria

Group C *Streptococci* were the most frequent isolated bacteria (27.0%), mainly from genital samples (56.4%) (Figure 1). No resistance was observed against penicillins and cephalosporins (Table 1). Between 2016 and 2019, the frequencies of Group C *Streptococci* highly resistant to streptomycin HC and kanamycin HC have significantly decreased over time, from 5.5% to 0.5% ($p < 0.0001$) and from 5.3% to no resistant strains ($p < 0.0001$), respectively. Despite an increased resistance to macrolides, rifampicin and sulphonamides observed in 2017 compared to 2016, the level of resistance decreased significantly in subsequent years. Concerning tetracycline, more than 82% of streptococcal isolates were resistant in 2016 and 2017, with a significant reduction to near 72% in 2018 and 60% in 2019 (Table 1).

Table 1. Percentage of resistant group C *Streptoccoci* isolates per year.

Antibiotic Category		Year (Number of Strains)	2016 (692)	2017 (598)	2018 (454)	2019 (374)
Penicillins		PEN	0.1	0.3	0	0
		AMX ** ($p = 0.016$)	0.7	0.2	0	0
		OXA	0.1	0.3	0	0
		AMC ** ($p = 0.011$)	0.7	0.0 * ($p = 0.037$)	0	0
Cephalosporins	3rd	CEF ** ($p = 0.049$)	0.4	0	0	0
	4th	CEQ	0.1	0	0	0
Aminoglycosides		STR ^{HC} ** ($p < 0.0001$)	5.5	3.8	0.0 *	0.5
		KAN ^{HC} ** ($p < 0.0001$)	5.3	4.5	0.2 * ($p < 0.0001$)	0
		GEN^{HC}	0.6	1.2	0.2	0
Tetracycline		TET ** ($p < 0.0001$)	82.1	87.0 * ($p < 0.0001$)	71.6 * ($p < 0.0001$)	58.6 * ($p < 0.0001$)
Macrolides		ERY ** ($p < 0.0001$)	11.1	22.1* ($p < 0.0001$)	10.3 * ($p < 0.0001$)	2.1 * ($p < 0.0001$)
Rifampicin		RIF	15.5	47.8 * ($p < 0.0001$)	22.0 * ($p < 0.0001$)	16.6 * ($p = 0.049$)
Sulphonamides		SXT ** ($p < 0.0001$)	4.8	15.6 * ($p < 0.0001$)	0.7 * ($p < 0.0001$)	0

PEN—penicillin; AMX—amoxicillin; AMC—amoxicillin-clavulanic acid; CEF—ceftiofur; CEQ—cefquinome; STR—streptomycin; KAN—kanamycin; GEN—gentamicin TET—tetracycline; ERY—erythromycin; RIF—rifampicin; SXT—trimethoprim-sulfamethoxazole; OXA—oxacillin, marker of methicillin resistance; * Chi-square test compared to the previous year, $p < 0.05$; ** Cochran–Armitage trend test, $p < 0.05$; ^{HC}—High concentration. The percentage of resistance is categorised by cell colours: green for resistance ≤10%, yellow for resistance between 10% and 30%, pink for resistance between 30% and 50%, and red for resistance >50%.

Most of these tetracycline-resistant group C *Streptococci* isolates were also of genital origin. They represented 83.5% in 2016, 90.4% in 2017, 82.1% in 2018 and 64.8% in 2019 (Figure S1a).

Staphylococcus aureus (*S. aureus*) is mainly isolated from cutaneous (48.8%) and genital samples (23.9%) (Figure 1). As shown in Table 2, a significant overall increase in the resistance to all antimicrobial drugs—except penicillin, amoxicillin, cefoxitin, streptomycin and erythromycin—used against *S. aureus* was observed, especially for aminoglycosides and tetracycline for which the percentage of resistant strains was more than 40% in 2019. The level of *S. aureus* resistant to oxacillin and cefoxitin (used as markers for methicillin resistance and extended to other β-lactams) increased from 15.8% to 28% over time.

Table 2. Percentage of resistant *Staphylococcus aureus* per year.

Antibiotic Category		Year (Number of Strains)	2016 (139)	2017 (118)	2018 (118)	2019 (107)
Penicillins		PEN	43.9	47.5	47.5	56.1
		AMX	43.2	46.6	47.5	55.1
		OXA ** ($p = 0.045$)	15.8	22.9	22.0	27.1
		AMC ** ($p = 0.021$)	17.3	22.9	22.6	28.0
Cephalosporins	2nd	FOX	17.3	22.9	22.6	28.0
	3rd	CEF ** ($p = 0.045$)	17.3	22.9	22.6	28.0
	4th	CEQ ** ($p = 0.031$)	17.3	22.9	22.6	28.0
Aminoglycosides		STR	20.9	11.0 * ($p = 0.033$)	11.0	17.8
		KAN ** ($p = 0.003$)	23.0	31.4	32.2	41.1
		GEN ** ($p = 0.001$)	21.6	30.5	32.2	41.1
Tetracycline		TET ** ($p = 0.01$)	27.3	35.6	35.6	43.9
Macrolides		ERY	5.8	5.1	4.2	8.4
Rifampicin		RIF ** ($p < 0.001$)	2.9	11.0 * ($p = 0.009$)	15.3	16.8

Table 2. *Cont.*

Antibiotic Category	Year (Number of Strains)	2016 (139)	2017 (118)	2018 (118)	2019 (107)
Sulphonamides	SXT ** ($p = 0.008$)	6.5	3.4	12.7 * ($p = 0.008$)	14.0
Fluoroquinolones	ENO ** ($p = 0.002$)	1.4	2.5	5.9	9.3
	MAR ** ($p = 0.002$)	1.4	1.7	5.9	9.3

PEN—penicillin; AMX—amoxicillin; AMC—amoxicillin-clavulanic acid; CEF—ceftiofur; CEQ—cefquinome; STR—streptomycin; KAN—kanamycin; GEN—gentamicin; TET—tetracycline; ERY—erythromycin; RIF—rifampicin; SXT—trimethoprim-sulfamethoxazole; ENO—enrofloxacin; MAR—marbofloxacine; OXA—oxacillin, marker of methicillin resistance; FOX—cefoxitin, marker of methicillin resistance; * Chi-square test compared to the previous year, $p < 0.05$; ** Cochran–Armitage trend test, $p < 0.05$. The percentage of resistance is categorised by cell colours: green for resistance ≤10%, yellow for resistance between 10% and 30%, pink for resistance between 30% and 50%, and red for resistance >50%.

3.2.2. GRAM Negative Bacteria

Escherichia coli (*E. coli*) was the second most frequent bacterium isolated (18%), mainly from genital samples (68.9%). Resistance to penicillins varied significantly during the 2016–2019 period, decreasing between 2016 and 2018, from 39.5% to 27.4%, for amoxicillin and from 31.4% to 18.3% for amoxicillin combined with clavulanic acid, and then increasing between 2018 and 2019, to 32.8% and 19.4%, respectively. Less than 6.2% of *E. coli* strains were resistant to cephalosporins (C3G and C4G) and quinolones. Concerning the resistance to streptomycin, after a significant decrease from 33.1% in 2016 to 26.2% in 2017 ($p = 0.048$), the highest percentage of resistant bacteria was measured in 2019, at 43.4% ($p < 0.0001$). For other aminoglycosides (such as kanamycin and gentamicin), resistant *E. coli* strains represented less than 11.2%. More than 76% and 67% of *E. coli* strains remained susceptible and sensitive to tetracycline and sulphonamides, respectively, and the percentage was 94.6% for quinolones (Table 3).

Table 3. Percentage of resistant *Escherichia coli* per year.

Antibiotic Category		Year (Number of Strains)	2016 (344)	2017 (325)	2018 (372)	2019 (341)
Penicillins		AMX ** ($p = 0.02$)	39.5	33.2	27.4	32.8
		AMC ** ($p < 0.0001$)	31.4	21.2 * ($p = 0.003$)	18.3	19.4
Cephalosporins	3rd	CEF	6.1	5.8	6.2	3.5
	4th	CEQ	5.8	5.8	6.2	3.8
Aminoglycosides		STR ** ($p = 0.005$)	33.1	26.2* ($p = 0.048$)	28.2	43.4 * ($p < 0.0001$)
		KAN	9.0	8.9	9.1	11.1
		GEN	6.1	7.1	8.9	6.7
Tetracycline		TET	20.6	21.2	23.1	22.6
Sulphonamides		SXT	31.4	28.3	28.8	32.6
Quinolones/Fluoroquinolones		NAL	4.9	3.4	5.4	3.2
		FLU	4.9	3.4	5.4	3.2
		ENO	3.2	3.4	2.4	3.2
		MAR	2.9	3.4	2.4	2.9

AMX—amoxicillin; AMC—amoxicillin-clavulanic acid; CEF—ceftiofur; CEQ—cefquinome; STR—streptomycin; KAN—kanamycin; GEN—gentamicin; TET—tetracycline; SXT—trimethoprim-sulfamethoxazole; ENO—enrofloxacin; MAR—marbofloxacine; FLU—flumequine; NAL—nalidixic acid, marker of fluoroquinolone resistance. * Chi-square test compared to the previous year, $p < 0.05$; ** Cochran–Armitage trend test, $p < 0.05$. The percentage of resistance is categorised by cell colours: green for resistance ≤10%, yellow for resistance between 10% and 30%, and pink for resistance between 30% and 50%.

The majority of resistant *E. coli* strains have been isolated from genital samples , where resistance to streptomycin increased from 28.4% in 2016 to 36.7% in 2019 (Figure S1c).

Pseudomonas aeruginosa (*P. aeruginosa*) represents 3.4% of the total bacteria and was mainly isolated from genital (46.6%) and respiratory (33.6%) samples (Figure 1). The analysis was limited to cefquinome (C4G), gentamicin and marbofloxacin, which are the only antimicrobials clinically relevant for *P. aeruginosa* in veterinary medicine. The frequency of strains resistant to these three agents was less than 15% during the 2016–2019 period (Table 4).

Table 4. Percentage of resistant *Pseudomonas aeruginosa* per year.

Antibiotic Category		Year (Number of Strains)	2016 (59)	2017 (70)	2018 (75)	2019 (64)
Cephalosporin	4th	CEQ	11.9	14.3	14.7	12.5
Aminoglycosides		GEN	10.2	8.6	14.7	10.9
Fluoroquinolones		MAR	1.7	0.0	0.0	4.7

CEQ—cefquinome; GEN—gentamicin; MAR—marbofloxacine. * Chi-square test compared to the previous year, $p < 0.05$; ** Cochran–Armitage trend test, $p < 0.05$. The percentage of resistance is categorised by cell colours: green for resistance ≤10% and yellow for resistance between 10% and 30%.

From 2018 to 2019, the frequency of strains resistant to cefquinome and gentamicin decreased, from 28.6% to 13.3% and from 32.1% to 20.0%, respectively, in respiratory samples (Figure S1d). Concerning resistant *P. aeruginosa* strains isolated from genital samples, a decrease was observed for cefquinome (from 26.1% in 2017 to 12.5% in 2019) and for gentamicin (from 13% in 2017 to 7.5% in 2019). These percentages have been calculated from the data in Table S1, and the variations are represented in terms of the numbers of isolates in Figure S1d.

Klebsiella pneumoniae (K. pneumoniae) represented 2.3% of all the bacteria and were mainly isolated from genital samples (63.9%) (Figure 1). The number of isolated strains doubled in four years from 30 to 60 strains (Table 5). A large increase in the resistance level to all antimicrobial agents was observed in 2017, especially for amoxicillin–clavulanic acid (from 12.9% to 42.4%), streptomycin (from 29.0% to 48.5%), tetracycline (from 25.8% to 48.5%) and sulphonamides (from 32.3% to 51.5%). Decreases were then measured in 2018 and confirmed in 2019, except for cephalosporins (change from 5.4% to 10%).

Table 5. Percentage of resistant *Klebsiella pneumoniae* per year.

Antibiotic Category		Year (Number of Strains)	2016 (31)	2017 (33)	2018 (56)	2019 (60)
Penicillins		AMC	12.9	42.4 * ($p = 0.009$)	16.1 * ($p = 0.006$)	10.0
Cephalosporins	3rd	CEF	9.7	21.2	5.4 * ($p = 0.022$)	10.0
	4th	CEQ	9.7	21.2	5.4 * ($p = 0.022$)	10.0
Aminoglycosides		STR ** ($p = 0.008$)	29.0	48.5	25.0 * ($p = 0.024$)	13.3
		KAN	3.2	12.1	7.1	6.7
		GEN	6.5	21.2	7.1	6.7
Tetracycline		TET ** ($p = 0.017$)	25.8	48.5	25.0 * ($p = 0.024$)	13.3
Sulphonamides		SXT ** ($p=0.006$)	32.3	51.5	26.8 * ($p = 0.019$)	15.0
Quinolones/Fluoroquinolones		NAL ** ($p = 0.049$)	19.4	21.2	8.9	8.3
		FLU	12.9	21.2	8.9	8.3
		ENO	9.7	18.2	3.6 * ($p = 0.02$)	5.0
		MAR	3.2	9.1 * ($p = 0.02$)	1.8	3.3

AMC—amoxicillin-clavulanic acid; CEF—ceftiofur; CEQ—cefquinome; STR—streptomycin; KAN—kanamycin; GEN—gentamicin; TET—tetracycline; SXT—trimethoprim-sulfamethoxazole; ENO—enrofloxacin; MAR—marbofloxacine; FLU—flumequine; NAL—nalidixic acid, marker of fluoroquinolone resistance. * Chi-square test compared to the previous year, $p < 0.05$; ** Cochran–Armitage trend test, $p < 0.05$. The percentage of resistance is categorised by cell colours: green for resistance ≤10%, yellow for resistance between 10% and 30%, pink for resistance between 30% and 50%, and red for resistance >50%.

Genital samples contained most of the resistant strains, with 63.9%. After an important increase in 2017, with 66.7% of genital strains resistant against tetracycline, 58.3% against amoxicillin with clavulanic acid and 50% against streptomycin and sulphonamides, the levels of strains resistant against all antimicrobial agents were below 10% in 2019 (Figure S1e).

Enterobacter spp represented 2.1% of total bacteria and was mainly isolated from genital samples (43.0%) (Figure 1). Only the evolution of strains resistant to streptomycin and kanamycin was significant over the 4-year period ($p = 0.022$ and $p = 0.044$ respectively). The percentage of *Enterobacter* spp. strains resistant to streptomycin increased significantly from 23.7% in 2016 to 50.0% in 2017 ($p = 0.025$) and to 55.2% in 2019. Similar evolution was observed for kanamycin ($p = 0.024$) and gentamicin ($p = 0.008$) resistance from 18.4% in 2016 to more than 41% in 2017. The frequency of strains resistant to cephalosporins varied between 15.8% and 34.6% for ceftiofur and between 10.3% and 21.7% for cefquinome. Strains resistant to tetracycline and sulphonamides increased from 21.1% to 37.9% and 48.3%, respectively. Concerning quinolone resistance, 27.6% of isolated strains were resistant to flumequine, 10.3% to enrofloxacin and 6.9% to marbofloxacine in 2019 (Table 6).

Table 6. Percentage of resistant *Enterobacter* spp. per year.

Antibiotic Category		Year (Number of Strains)	2016 (38)	2017 (46)	2018 (52)	2019 (29)
Cephalosporins	3rd	CEF	15.8	30.4	34.6	27.6
	4th	CEQ	13.2	21.7	21.2	10.3
Aminoglycosides		STR ** ($p = 0.022$)	23.7	50.0 * ($p = 0.025$)	44.2	55.2
		KAN ** ($p = 0.044$)	18.4	41.3 * ($p = 0.024$)	36.5	44.8
		GEN	18.4	45.7 * ($p = 0.008$)	42.3	41.4
Tetracycline		TET	21.1	32.6	36.5	37.9
Sulphonamides		SXT	21.1	45.7	42.3	48.3
Quinolones/Fluoroquinolones		NAL	21.1	17.4	23.1	27.6
		FLU	21.1	17.4	23.1	27.6
		ENO	7.9	8.7	13.5	10.3
		MAR	2.6	4.3	7.7	6.9

CEF—ceftiofur; CEQ—cefquinome; STR—streptomycin; KAN—kanamycin; GEN—gentamicin; TET—tetracycline; SXT—trimethoprim-sulfamethoxazole; ENO—enrofloxacin; MAR—marbofloxacine; FLU—flumequine; NAL—nalidixic acid, marker of fluoroquinolone resistance. * Chi-square test compared to the previous year, $p < 0.05$; ** Cochran–Armitage trend test, $p < 0.05$. The percentage of resistance is categorised by cell colours: green for resistance ≤10%, yellow for resistance between 10% and 30%, pink for resistance between 30% and 50%, and red for resistance >50%.

The distribution of resistant strains according to sample origins revealed that genital samples contained the largest number of resistant bacteria (43%). In 2019, 21.4 % of *Enterobacter* spp. isolated from the genital tract were resistant to cefquinome and tetracycline, 42.9 % to streptomycin, 35.7% to kanamycin, and 28.6% to gentamicin and sulphonamides (Figure S1f).

3.3. Multi-Drug Resistant (MDR) Bacteria

Table 7 shows the overall level and trend of MDR bacteria (defined as non-susceptible to at least three different classes of antibiotic usually efficient) according to bacterial species. Because few antimicrobial compounds have been tested against *P. aeruginosa*, this species was excluded from the analysis. For group C *Streptococci*, the level of MDR increased from 2016 (10.7%) to 2017 (18.9%) before decreasing to 0.5% in 2019. For *S. aureus*, the percentage of MDR increased from 24.5% in 2016 to 37.4% in 2019. For *E. coli*, the level of MDR remained similar during the 2016–2019 period (average of resistance was 22.0%). For *K. pneumoniae*, the level of MDR, after an increase from 38.7% in 2016 to 51.5% in 2017, decreased to 11.7% in 2019. Finally, for *Enterobacter* spp., this level doubled from 26.3% in 2016 to 51.7% in 2019.

Table 7. Percentage of bacteria resistant to three or more antimicrobial classes.

	Streptococcus (Group C) ** ($p < 0.001$)	*Staphylococcus aureus* ** ($p = 0.029$)	*E. coli*	*Klebsiella pneumoniae* ** ($p = 0.001$)	*Enterobacter* spp. ** ($p = 0.048$)
2016	10.7	24.5	22.7	38.7	26.3
2017	18.9	31.4	21.2	51.5	45.6
2018	3.1	33.1	21.8	26.8	44.2
2019	0.5	37.4	22.6	11.7	51.7

** Cochran–Armitage trend test, $p < 0.05$.

4. Discussion

Horses are now recognised to be potential reservoirs of antimicrobial resistance, which can be transmitted to other animal and human pathogens [11,12,14–16]. Equine pathogens with zoonotic potential should be carefully taken into account, especially *Enterobacteriaceae* producing extended-spectrum beta-lactamases (ESBL), *P. aeruginosa* or methicillin-resistant *Staphylococcus aureus* (MRSA), as described by the WHO [17].

Human infection with Group C *Streptococci* is not frequent but can lead to severe diseases, such as septicemia, meningitis or arthritis [18–20]. In horses, *Streptococcus equi* subsp *equi* and *Streptococcus equi* subsp *zooepidemicus* (*Streptococcus zooepidemicus*) are the most important bacterial pathogens encountered, the causative agent of Strangles and the leading cause of bacterial infection, respectively [21–23]. Only very few (less than 1%) strains resistant to penicillins and cephalosporins were identified in the 4-year study, in agreement with other studies conducted in France [11,12] and throughout the world [6–10]. However, a significantly increased resistance to tetracycline, macrolides, rifampicin and sulphonamides was measured in 2017, followed by a significant decrease in 2018 and 2019. These results were well correlated with the trend of MDR *Streptococci* that represented 0.5% in 2019. *Streptococcus zooepidemicus* is frequently associated with uterine infections but can also induce persistent subclinical infection of the mare due to the presence of "dormant" bacterial colonies in the endometrium, with an impact on fertility. This stage of dormancy is often associated with an increased resistance to penicillin (not linked to the acquisition of a resistance gene but due to the absence of replication) [24]. The instillation of a bacterial growth medium (bActivate) in the uterus has been shown to "reactivate" dormant *Streptococcus zooepidemicus*, which subsequently allows the efficient use of antibiotics to clear the persistent infection [24]. It is also important to note that *Streptococcus zooepidemicus* is often associated with secondary bacterial infections in horses after respiratory virus infections [25]. The use of vaccination against the primary pathogen should also be considered as an indirect way to reduce antibiotic use through the prevention of secondary bacterial infections, as previously demonstrated for equine influenza vaccinations [26].

Concerning *S. aureus*, an increased frequency of MRSA strains was observed between 2016 and 2019. As shown in Table 2, the level of oxacillin-resistant strains (used as a methicillin resistance marker) increased by more than 10% (from 15.8% in 2016 to 27.1% in 2019). More strains resistant against aminoglycosides and tetracyclines were also isolated. In parallel, we showed that the level of MDR *S. aureus* also increased during the same period. As the level of MRSA remained stable between 2013 and 2016, the current result raises questions regarding long-term vigilance and the correct use of antimicrobial compounds [11]. Moreover, in a previous study, we have demonstrated the predominance of ST398 MRSA isolates since 2011. Since this ST398 type is known to cause outbreaks in horses and to colonise/infect humans, hygiene measures and appropriate antimicrobial use should be maintained and reinforced in order to limit the transmission of *S. aureus* between horses as well as between horses and humans [27].

Concerning *E. coli*, resistance to cephalosporins and quinolones represented, in 2019, less than 10% of all isolated strains, as observed in previous years [11,12], which is well correlated with their limited use in the field because of their critical categorisation by authorities. Indeed, one study that measured the trends in the antimicrobial susceptibility of several pathogens, such as *Enterobacteriacae*, isolated between 1979 and 2010 from foals with sepsis highlighted a decrease in ceftiofur activity [28]. For other drugs, the resistance levels measured in this study were similar to the levels reported before 2016, with the exception of those for streptomycin, which reached 43.4% in 2019. Even if it was significantly reduced two-fold compared to 2009 and 2010, where 94.8% and 90.4% were resistant, respectively, this percentage remained high between 2016 and 2019 (26.2% in 2017 and 43.4% in 2019). This may be explained by the fact that *E. coli* strains were mainly isolated from the genital tract and that streptomycin represents the main treatment for such infections, thereby promoting the selection of streptomycin (STR)-resistant *E. coli* isolates. This again points to the necessity of performing antimicrobial susceptibility testing before any antimicrobial treatment.

In a lower proportion than *E. coli*, two other Enterobacteria were also analysed, *K. pneumoniae* (2.3% of isolated bacteria) and *Enterobacter spp* (2.1% of isolated bacteria). For *K. pneumoniae*, unexpectedly, the frequency of strains resistant against all antimicrobial agents increased in 2017 before reaching levels of resistance similar to those observed during 2006–2016 period. For *Enterobacter* spp., the same observation was made in 2017, with a subsequent maintenance of high levels of resistance against aminoglycosides, streptomycin and kanamycin especially, until 2019. The MDR proportion of *Enterobacter* spp. also doubled between 2016 and 2019. As mentioned above, the increased resistance of *K. pneumoniae* and *Enterobacter* spp. observed in 2017 could be linked to the transition between the two French national programmes leading to reduced vigilance regarding the antimicrobials used. However, although *Enterobacter* spp. represented the smallest sample of pathogens in this study (2.1%), attention must be paid to multi-drug resistant strains whose levels remained very high until 2019. Marbofloxacine appeared to be the compound against which few *Enterobacter* were resistant, and it should be used with caution to avoid a therapeutic impasse.

As described in literature [9,11,29], only a few antimicrobial agents such as cefquinome, gentamicin and marbofloxacine remained active against *P. aeruginosa*. Moderate levels of resistance (around 8.6%, 14.7% and 4.7% for cefquinome, gentamicin and marbofloxacine, respectively) were measured between 2016 and 2019, as already observed in 2015 [11]. *Pseudomonas aeruginosa* strains were mainly isolated from the respiratory and genital tracts, where antimicrobial drugs are more targeted because of reproduction and horse racing, inducing less selective pressure.

Overall, the antimicrobial resistance of the bacteria studied here increased in 2017, which was the pivotal year between the two governmental programmes implemented in France to counteract antimicrobial resistance. A decrease was subsequently observed. Nevertheless, high levels of MDR persist as of 2019, especially in more than 37% of *S. aureus* and 51% of *Enterobacter* spp. This situation will have to be carefully monitored in the future. The use of these antimicrobials has to be moderated in order to prevent the spread of resistance.

The majority of the bacterial collection analysed in this study was isolated from the genital tract, which is linked to the primary diagnostic activity of the LABÉO Veterinary Microbiology unit. Consequently, this study may lack representativeness regarding other compartments.To date, the use of antimicrobials in equine veterinary medicine remains a necessity in many cases because the methods for the prevention of bacterial infections in horses are limited, with few or no vaccines available.

In this context, antimicrobials are essential to equine health, and this study underlines the importance of performing antimicrobial susceptibility testing in order to optimise antimicrobial therapy in horses and to reduce the occurrence of resistance. Such studies are essential for evaluating the evolution of antimicrobial resistance and its potential threat to public health. In addition, such data are key indicators for the impact of national or international plans leading to a reduction in the spread of multi-drug resistant bacteria and may help in drawing up new guidelines. The analysis of the resistance mechanisms displayed by these major bacteria is warranted. The use of whole genome sequencing of strains of interest will prove to be invaluable for investigating molecular epidemiology.

5. Conclusions

During the 2016-2019 period, decreases in the resistance of group C Streptococci and Klebsiella pneumoniae against at least four classes of antimicrobials, were observed. In addition, Staphylococcus aureus, and Enterobacter spp. presented an increased resistance against all the classes tested. In this context, the percentages of multi-drug resistant strains of these species increased from 24.5% to 37.4% and from 26.3% to 51.7%, respectively. For E. coli, the situation was mixed with a decrease against penicillins and an increase against streptomycin and sulphonamides. Our study indicates that horses may be considerate as reservoirs of antimicrobial resistant pathogens and underlines the importance of performing antimicrobial susceptibility testing in order to optimize antimicrobial therapy in horses and to reduce the occurrence of these resistances.

Supplementary Materials: The following are available online at http://www.mdpi.com/2076-2615/10/5/812/s1, Table S1: Number of resistant strains of the major bacterial species isolated over a 4-years period (2016–2019) according to the origin. R-respiratory; D-digestive; C-cutaneous; G-genital and O-other, Figure S1: Total (blue) and resistant (red) relevant Gram-positive (a, b) and Gram-negative (c, d, e, f) bacteria isolated per year and major sample groups.

Author Contributions: Conceptualization, A.L. and J.-C.G.; methodology, K.M. and A.L.; software, A.L. and S.C.; validation, K.M., R.P. and J.-C.G.; formal analysis, A.L. and S.C.; investigation, A.L.; resources, K.M. and A.L.; data curation, K.M., S.C. and A.L.; writing—original draft preparation, A.L.; writing—review and editing, S.C., K.M., R.P. and J.-C.G.; visualization, A.L.; supervision, A.L.; project administration, A.L. All authors have read and agreed to the published version of the manuscript.

Funding: This research received no external funding.

Acknowledgments: We would like to thank the technical staff at the Veterinary Microbiology unit of LABÉO Frank Duncombe for the achievement of antimicrobial susceptibility tests.

Conflicts of Interest: The authors declare no conflict of interest.

References

1. World Health Organization and Antimicrobial Resistance. Available online: https://www.who.int/health-topics/antimicrobial-resistance (accessed on 16 March 2020).

2. Food and Agriculture Organisation of the United Nations and Antimicrobial Resistance. Available online: http://www.fao.org/antimicrobial-resistance/fr/ (accessed on 16 March 2020).

3. World Organisation for Animal Health-OIE and Antimicrobial Resistance. Available online: https://www.oie.int/en/for-the-media/amr/ (accessed on 16 March 2020).

4. The ECOANTIBIO Plan 2012–2017. Available online: https://agriculture.gouv.fr/plan-ecoantibio-2012-2017-lutte-contre-lantibioresistance (accessed on 25 March 2020).

5. ECOANTIBIO2. Available online: https://agriculture.gouv.fr/le-plan-ecoantibio-2-2017-2021 (accessed on 25 March 2020).

6. Chipangura, J.K.; Chetty, T.; Kgoete, M.; Naidoo, V. Prevalence of antimicrobial resistance from bacterial culture and susceptibility record from horse samples in South Africa. *Prev. Vet. Med.* **2017**, *148*, 37–43. [CrossRef] [PubMed]

7. Clark, C.; Greenwood, S.; Boison, J.O.; Chirino-Trejo, M.; Dowling, P.M. Bacterial isolates from equine infections in western Canada (1998–2003). *Can. Vet. J.* **2008**, *49*, 153–160. [PubMed]

8. Malo, A.; Cluzel, C.; Labrecque, O.; Beauchamp, G.; Lavoie, J.P.; Leclere, M. Evolution of *in vitro* antimicrobial resistance in an equine hospital over 3 decades. *Can. Vet. J.* **2016**, *57*, 747–751.

9. van Spijk, J.N.; Schmitt, S.; Fürst, A.E.; Schoster, A. A retrospective analysis of antimicrobial resistance in bacterial pathogens in an equine hospital (2012–2015). *Schweiz. Arch. Tierheilkd.* **2016**, *158*, 433–442. [CrossRef]

10. Johns, I.C.; Adams, E.L. Trends in antimicrobial resistance in equine bacterial isolates: 1999–2012. *Vet. Rec.* **2015**, *176*, 334. [CrossRef]

11. Duchesne, R.; Castagnet, S.; Maillard, K.; Petry, S.; Cattoir, V.; Giard, J.C.; Leon, A. In vitro antimicrobial susceptibility of equine clinical isolates from France, 2006–2016. *J. Glob. Antimicrob. Resist.* **2019**, *19*, 144–153. [CrossRef]

12. Bourély, C.; Cazeau, G.; Jarrige, N.; Haenni, M.; Gay, E.; Leblond, A. Antimicrobial resistance in bacteria isolated from diseased horses in France. *Equine Vet. J.* **2020**, *52*, 112–119. [CrossRef]

13. CA-SFM/EUCAST. Available online: https://www.sfm-microbiologie.org/2019/05/06/casfm-eucast-2019-v2/ (accessed on 3 April 2020).

14. Schmiedel, J.; Falgenhauer, L.; Domann, E.; Bauerfeind, R.; Prenger-Berninghoff, E.; Imirzalioglu, C.; Chakraborty, T. Multiresistant extended-spectrum β-lactamase-producing Enterobacteriaceae from humans, companion animals and horses in central Hesse, Germany. *BMC Microbiol.* **2014**, *12*, 14–187. [CrossRef]

15. EMA/CVMP/AWP/401740/2013. Reflection Paper on the Risk of Antimicrobial Resistance Transfer from Companion Animals. Available online: https://www.ema.europa.eu/en/documents/scientific-guideline/reflection-paper-risk-antimicrobial-resistance-transfer-companion-animals_en.pdf (accessed on 25 March 2020).

16. Argudín, M.A.; Deplano, A.; Meghraoui, A.; Dodémont, M.; Heinrichs, A.; Denis, O.; Nonhoff, C.; Roisin, S. Bacteria from Animals as a Pool of Antimicrobial Resistance Genes. *Antibiotics* **2017**, *6*, 12. [CrossRef] [PubMed]

17. WHO. Global Priority List of Antibiotic-Resistant Bacteria to Guide Research, Discovery, and Development of New Antibiotics. 2017. Available online: https://www.who.int/medicines/publications/global-priority-list-antibiotic-resistant-bacteria/en/ (accessed on 26 March 2020).

18. Ferrandière, M.; Cattier, B.; Dequin, P.F. Septicemia and meningitis due to *Streptococcus zooepidemicus*. *Eur. J. Clin. Microbiol. Infect. Dis.* **1998**, *17*, 290–291. [CrossRef] [PubMed]

19. Friederichs, J.; Hungerer, S.; Werle, R.; Militz, M.; Bühren, W. Human bacterial arthritis caused by *Streptococcus zooepidemicus*: Report of a case. *Int. J. Inf. Dis.* **2010**, e233–e235. [CrossRef] [PubMed]

20. Björnsdóttir, S.; Harris, S.R.; Svansson, V.; Gunnarsson, E.; Gammeljord, K.; Steward, K.F.; Newton, J.R.; Robinson, C.; Charbonneau, A.R.; Parkhill, J.; et al. Genomic dissection of an Icelandic epidemic of respiratory disease in horses and associated zoonotic cases. *mBio* **2017**, *8*, e00826-17. [CrossRef]

21. Waller, A.S. Strangles: A pathogenic legacy of the war horse. *Vet. Rec.* **2016**, *178*, 91–92. [CrossRef] [PubMed]

22. Waller, A.S. Science-in-brief: *Streptococcus zooepidemicus*: A versatile opportunistic pathogen that hedges its bets in horses. *Equine Vet. J.* **2017**, *49*, 146–148. [CrossRef] [PubMed]

23. Boyle, A.G.; Timoney, J.F.; Newton, J.R.; Hines, M.T.; Waller, A.S.; Buchanan, B.R. *Streptococcus equi* Infections in Horses: Guidelines for Treatment, Control, and Prevention of Strangles-Revised Consensus Statement. *J. Vet. Intern Med.* **2018**, *32*, 633–647. [CrossRef]

24. Petersen, M.R.; Skive, B.; Christoffersen, M.; Lu, K.; Nielsen, J.M.; Troedsson, M.H.; Bojesen, A.M. Activation of persistent *Streptococcus equi* subspecies *zooepidemicus* in mares with subclinical endometritis. *Vet. Microbiol.* **2015**, *179*, 119–125. [CrossRef]

25. Muranaka, M.; Yamanaka, T.; Katayama, Y.; Niwa, H.; Oku, K.; Matsumura, T.; Oyamada, T. Time-related Pathological Changes in Horses Experimentally Inoculated with Equine Influenza A Virus. *J. Eq. Sci.* **2012**, *23*, 17–26. [CrossRef]

26. Paillot, R.; Prowse, L.; Montesso, F.; Huang, C.M.; Barnes, H.; Escala, J. Whole inactivated equine influenza vaccine: Efficacy against a representative clade 2 equine influenza virus, IFNgamma synthesis and duration of humoral immunity. *Vet. Microbiol.* **2013**, *162*, 396–407. [CrossRef]

27. Guérin, F.; Fines-Guyon, M.; Meignen, P.; Delente, G.; Fondrinier, C.; Bourdon, N.; Cattoir, V.; Léon, A. Nationwide molecular epidemiology of methicillin-resistant *Staphylococcus aureus* responsible for horse infections in France. *BMC Microbiol.* **2017**, *17*, 104. [CrossRef]

28. Theelen, M.J.; Wilson, W.D.; Edman, J.M.; Magdesian, K.G.; Kass, P.H. Temporal trends in in vitro antimicrobial susceptibility patterns of bacteria isolated from foals with sepsis: 1979-2010. *Equine Vet. J.* **2014**, *46*, 161–168. [CrossRef]

29. Davis, H.A.; Stanton, M.B.; Thungrat, K.; Boothe, D.M. Uterine bacterial isolates from mares and their resistance to antimicrobials: 8296 cases (2003–2008). *J. Am. Vet. Med. Assoc.* **2013**, *242*, 977–983. [CrossRef] [PubMed]

Article

Prevalence, Risk Factors, and Characterization of Multidrug Resistant and ESBL/AmpC Producing *Escherichia coli* in Healthy Horses in Quebec, Canada, in 2015–2016

Maud de Lagarde [1], John M. Fairbrother [1,*] and Julie Arsenault [2]

[1] OIE Reference Laboratory for *Escherichia coli*, Faculty of Veterinary Medicine, Université de Montréal, Saint-Hyacinthe, QC J2S2M2, Canada; maud.de.lagarde@umontreal.ca

[2] Epidemiology of Zoonoses and Public Health Research Unit (GREZOSP), Faculty of Veterinary Medicine, Université de Montréal, Saint-Hyacinthe, QC J2S2M2, Canada; julie.arsenault@umontreal.ca

* Correspondence: john.morris.fairbrother@umontreal.ca

Received: 20 January 2020; Accepted: 19 March 2020; Published: 20 March 2020

Simple Summary: Antimicrobial resistance has been recognised as a global threat by the WHO. ESBL/AmpC genes, responsible for cephalosporin resistance, are particularly worrisome. *Escherichia coli* is a ubiquitous bacterium. Most strains are commensal, although some can cause disease in humans and animals. Due to its genome plasticity, it is a perfect candidate to acquire resistance genes. We hypothesized that multidrug-resistant *E. coli* and *E. coli* resistant to cephalosporins are present in the fecal microbiota of healthy horses in Quebec. We characterised antimicrobial resistance, identified ESBL/AmpC genes and assessed potential risk factors for their presence. Fecal samples from 225 horses, distributed in 32 premises, were cultured for indicator *E. coli* (selected without enrichment) and specific *E. coli* (selected after enrichment with ceftriaxone). Of the 209 healthy horses in which *E. coli* were detected, 46.3% shed multidrug-resistant (resistant to three or more classes of antimicrobials tested) *E. coli*. Non-susceptibility was most frequently observed for ampicillin, amoxicillin/clavulanic acid or streptomycin. ESBL/AmpC genes were detected in *E. coli* from 7.3% of horses and 18.8% of premises. The number of staff and equestrian event participation within the last three months were identified as risk factors for horses shedding multidrug-resistant *E. coli* isolates. The horse intestinal microbiota is a reservoir for ESBL/AmpC genes. The presence of ESBL/AmpC in horses is both a public and equine health concern, considering the close contact between horses and owners.

Abstract: Although antimicrobial resistance is an increasing threat in equine medicine, molecular and epidemiological data remain limited in North America. We assessed the prevalence of, and risk factors for, shedding multidrug-resistant (MDR) and extended-spectrum β-lactamase (ESBL) and/or AmpC β-lactamase-producing *E. coli* in healthy horses in Quebec, Canada. We collected fecal samples in 225 healthy adult horses from 32 premises. A questionnaire on facility management and horse medical history was completed for each horse. Indicator (without enrichment) and specific (following enrichment with ceftriaxone) *E. coli* were isolated and tested for antimicrobial susceptibility. The presence of ESBL/AmpC genes was determined by PCR. The prevalence of isolates that were non-susceptible to antimicrobials and to antimicrobial classes were estimated at the horse and the premises level. Multivariable logistic regression was used to assess potential risk factors for MDR and ESBL/AmpC isolates. The shedding of MDR *E. coli* was detected in 46.3% of horses. Non-susceptibility was most commonly observed to ampicillin, amoxicillin/clavulanic acid or streptomycin. ESBL/AmpC producing isolates were detected in 7.3% of horses. The most commonly identified ESBL/AmpC gene was $bla_{CTX-M-1}$, although we also identified bla_{CMY-2}. The number of staff and equestrian event participation were identified as risk factors for shedding MDR isolates. The prevalence of healthy horses harboring MDR or ESBL/AmpC genes isolates in their intestinal microbiota is noteworthy. We identified risk factors which could help to develop guidelines to preclude their spread.

Keywords: antimicrobial resistance; beta-lactamase; cephalosporinase; microbiota; North America; equine

---

1. Introduction

Antimicrobial resistance was reported by the World Health Organization (WHO) in 2014 as the largest current threat for global health [1]. Equine medicine is also involved, indeed, the first bacteria resistant to antimicrobials in horses were reported in 1971, in Canada [2]. Subsequently, the number of treatment failure reports due to antimicrobial resistance has increased [3–5]. In Europe, several studies have reported that healthy horses can carry multidrug resistant (MDR) bacteria at a relatively high prevalence (39% to 44%) [6,7] and some countries are setting up surveillance monitoring [8]. Nonetheless, molecular and epidemiological data in this species are still limited in North America. In the global approach to antimicrobial resistance recommended by the WHO, horses have been classified as companion animals, although they are also working animals and livestock and could contaminate their owner through direct contact, or even the general population via the food chain. Thus, horses have been overlooked in the general approach to antimicrobial resistance [9].

Escherichia coli is ubiquitous and mainly commensal in the intestinal microbiota of mammals. However, pathogenic strains have been recognized, mostly in human and in food-producing animals, and occasionally in horses [10]. Due to its ubiquity, recurrent exposure to systemic (oral, intramuscular or intravenous) antimicrobial treatment and the fast evolution of its genome, this bacterium is considered by the Canadian Integrated Program for Integrated Surveillance System (CIPARS) as an excellent indicator for antimicrobial resistance surveillance [11].

One of the main mechanisms of resistance in *E. coli* is the production of extended spectrum β-lactamases (ESBL) and/or AmpC cephalosporinases (AmpC) [12], resulting in the hydrolyzation of the β-lactam ring, which is present in penicillins, cephalosporins and monobactams. β-lactamase genes (*bla*) have spread very effectively among numerous species of Gram-negative bacteria over the last 30 years [12], both in animals and in humans. In horses, phenotypic resistance to ceftiofur, a third-generation cephalosporin, has been reported in many clinical situations [13]. $bla_{CTX-M-1}$ is the ESBL resistance gene variant most often detected [14]. However, other variants of CTX-M (i.e., $bla_{CTX-M-2}$, $bla_{CTX-M-9}$, $bla_{CTX-M-14}$, $bla_{CTX-M-15}$) have been recognized. bla_{CMY} and bla_{SHV-12} have also been identified [14]. All of these variants have also been found in other animal species [14] and in humans [15]. These genes spread mainly through plasmids, carrying multiple resistance genes. Thus, these plasmids convey resistance to other antimicrobial classes, promoting multidrug resistance [16]. However, the resistance gene dissemination can also be enhanced through "high-risk" clones [17]. An example of such a clone is the *E. coli* sequence-type ST410 [18], recently emerging as a public health concern in the human population.

Moreover, owning a horse has been demonstrated as a risk factor for the carriage of ESBL in people [19] in the Netherlands. Even though the author of this study nuanced these results by stating that horse owners often own other pets, and the Netherlands has a high population density which might not be representative of the situation of other countries, nevertheless, this study underlines the potential concern for human health. The colonization with ESBL-producing Enterobacteriaceae, in humans, has been associated with an increase in the length of hospitalization in ICU patients [20]. New regulations restricting the use of antimicrobials such as fluoroquinolones and cephalosporins, classified as having the highest priority by the WHO and Health Canada [21,22], to cases where the veterinarian can prove that there is no better alternative [23], came into effect in early 2019 in veterinary medicine in Canada. Nevertheless, the use of ceftiofur will likely remain common in horses due to the lack of a better alternative, especially for neonatal sepsis and respiratory diseases in adults, possibly enhancing the dissemination of ESBL/AmpC genes.

No data are available on the presence of MDR or ESBL/AmpC-producing isolates in the healthy equine population in Quebec. Our objective was to estimate the prevalence of, and risk factors for, shedding MDR- and/or ESBL/AmpC-producing *E. coli* isolates in horses. We characterized potential ESBL/AmpC isolates for antimicrobial susceptibility and the presence of ESBL/AmpC-associated resistance genes.

2. Materials and Methods

2.1. Sampling and Data Collection

During the summer 2015, MDL sampled healthy horses from a convenience sample of premises owned by clients and located within a one-hour drive from the CHUV, a university veterinary hospital located in Saint Hyacinthe, Quebec, Canada. To increase the number of sampled horses, in April 2016, the 111 Quebec association of equine veterinary practitioners (AVEQ) members were invited to a conference introducing the project. This event took place in Saint Hyacinthe, Quebec, and was also given in a videoconference. The veterinarians were solicited to sample healthy horses in the stables they visited as part of their veterinary practice. To evaluate the number of targeted horses sampled we used the following equation

$$n = (Z^2 \times P(1 - P))/L^2$$

where n = the number of targeted horses (n = 359 horses), Z = the value from the normal distribution corresponding to the 95% confidence interval (Z = 1.96), and P = the expected prevalence of MDR *E. coli* in the healthy horse population which we extrapolated from a previous article in Great Britain [6] (P = 0.37), and L the desired precision (L = 0.05). Given the large size of the total horse population in Quebec (estimated at 129,000 individuals by Equine Canada in 2010), we have not adjusted the number of horses for a finite population. This figure does not consider the potential non-independence of the status of horses in the same premises.

Every participating veterinarian receive d a sampling kit, including 100 rectal swabs (BBL^{TM} CultureSwab^{TM} Plus, Becton Dickinson, France) and the material to ship the samples to the Ecl laboratory at 4 °C, within 48 h of collection. Protocols were explained in detail in the kit. Each veterinarian could sample up to 10 horses per premise up to a maximum of 10 premises, until the overall target sample size was reached. Only horses over two years old and considered healthy by their owner were eligible for the study. We focused our study on adult horses because breeding does not take a huge place in equestrian activity in Quebec (around 1% of horse riders are interested in breeding in Quebec according to the Cheval Quebec activity report in 2016 (https://cheval.quebec/Rapport-annuel)), therefore we expect that most contacts between people and horses during these activities are with adult horses. The sampled horses were not necessarily part of the veterinarian clientele. Each owner agreed to participate on a voluntary basis. The protocol was approved by the Université de Montréal Ethic Committee for use of animals (15-Rech-1800).

For each sampled horse, the owner and the recruiting veterinarian each completed a questionnaire online, through the Surveymonkey web platform (https://www.surveymonkey.com). They were available in both French and English and are found in the Supplementary Data of this article (Supplementary Material Figures S1 and S2). Questions were based on previously reported and suspected risk factors in the horse [24] and were related to the facility management and horse medical history. Each premise was geocoded based on its 6-digit postal code, performed in GeoPinpoint suite version 6.4 (DMTI Spatial).

2.2. Indicator Collection: Non-enriched Culture, Antimicrobial Susceptibility Testing, ESBL/AmpC Gene Identification, and Prevalence Estimation

On reception at the Ecl Laboratory, rectal swabs were held in Luria-Bertani (LB) broth for a maximum of 15 min at room temperature, then 100 µL of LB broth was transferred on MacConkey agar and incubated at 37 °C overnight. All lactose-positive colonies, up to a maximum of three, were selected for each rectal sample and cultured in LB broth then plated on MacConkey agar to ensure purity. Isolates were confirmed as *E. coli* by the presence of the *uidA* gene [25], as detected by PCR. Each sample and isolate were stored in 15% glycerol at −80 °C.

Isolates were tested for susceptibility to the 14 antimicrobial agents examined in the CIPARS using the disk-diffusion assay. We used the same disks and techniques as described for the indicator collection of our previous work [7].

When isolates were non-susceptible (intermediate or resistant) to 3rd generation cephalosporins, we looked for 5 β-lactamase resistance genes (bla_{SHV}, bla_{TEM}, bla_{CMY-2}, bla_{OXA}, bla_{CTX-M}) by multiplex PCR. We used the same DNA extraction, PCR protocols and CTX-M-variant identification protocols as described for the indicator collection of our previous work [7].

We estimated the prevalence and 95% confidence intervals of (1) horses shedding non-susceptible (i.e., resistant or intermediate) isolate(s) for each antimicrobial, and (2) horses shedding isolate(s) non-susceptible to ≥ 1, 3, 5, 7 and 9 classes of antimicrobials. An isolate was considered MDR if non-susceptible to at least one agent in three or more antimicrobial classes [26]. We used the same method of calculation (with adjustment for sampling weights and clustering within premises) and the same software as described for in the indicator collection of our previous work [7]. We also estimated these prevalences and 95% confidence intervals at the premises level, as previously described [7]; for each outcome, a positive status was attributed when the premises housed at least one positive horse.

2.3. Potential ESBL/AmpC Producing E. coli Collection: Culture, Antimicrobial Susceptibility Testing, ESBL/AmpC and Virulence Gene Identification and Descriptive Statistics

ESBL/AmpC-producing bacteria may be shed in small quantities in healthy individuals [27]. To improve detection sensitivity and allow for a more accurate estimation of the proportion of positive horses, we carried out enrichment with ceftriaxone [8,27]. For each rectal swab suspension in LB broth, 1 mL was inoculated in 9 mL of MacConkey broth containing 1 mg/L of ceftriaxone and incubated overnight at 37 °C. When bacterial growth was positive, 100 μL of MacConkey broth was inoculated on MacConkey agar containing 1 mg/L of ceftriaxone and incubated at 37 °C overnight. All isolates up to a maximum of five lactose-positive isolates per sample were selected. All isolates of this collection were confirmed as *E. coli* by the presence of the *uidA* gene [25] as detected by PCR and were tested for susceptibility to 14 antimicrobials, as described above. All isolates in this collection were systematically examined for the presence of five β-lactamase resistance genes (bla_{SHV}, bla_{TEM}, bla_{CMY-2}, bla_{OXA}, bla_{CTX-M}) by multiplex PCR (PCR and gene identification protocols are described above).

Descriptive statistics were used to present the non-susceptibility patterns of isolates from this collection. We estimated prevalence with 95% confidence intervals of horses shedding ESBL/AmpC isolates and of the premises housing these horses, using the same calculation method described above.

2.4. MDR and ESBL/AmpC: Risk Factors

For the risk factor analyses, two outcome variables were investigated: MDR and ESBL/AmpC status for each horse. A positive MDR or ESBL/AmpC status was defined as the detection of at least one MDR or ESBL/AmpC isolate, respectively, for that horse. All potential risk factors from the questionnaire were categorized. Putative risk factors with $p < 0.20$ (Wald test) in univariable multilevel (facilities, horses) logistic regressions were selected for inclusion in a full multivariable multilevel model for each outcome. Pairwise associations between these selected variables were assessed by χ^2 test; in the presence of significant association ($p < 0.05$), only one of two correlated variables was kept based on the biological relevance with the outcome. The final multivariable model was refined by sequentially omitting variables with $p > 0.05$ (Wald test). Analyses were performed in MLwiN version 2.36 using 2nd order penalized quasi-likelihood estimation, with no extrabinomial variation permitted. The fit of the final model was evaluated by visual assessment of standardized residuals at the premise level against normal scores and against fixed part prediction.

3. Results

In 2015, MDL sampled 67 horses distributed in 10 premises. Following the conference, in April 2016, 14 equine practitioners agreed to participate in the study. Although samples were collected one year apart, the results are presented together as the sampling was similar and there was no modification in the equine practice in Quebec from 2015 to 2016.

A total of 225 horses were sampled, distributed in 32 premises, as illustrated in Figure 1. Between two and 12 horses were sampled in each premise, with a mean of seven horses sampled.

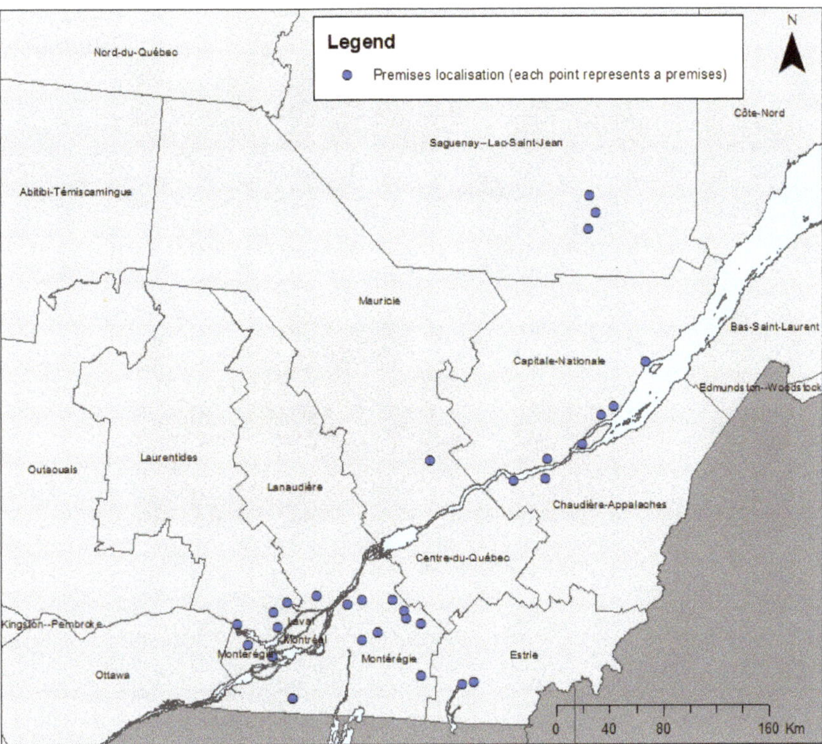

Figure 1. Geographical distribution of the sampled premises (based on center point of their 6-digit postal code) over the administrative regions of the province of Quebec in a cross-sectional study of 209 healthy adult horses in 32 premises performed in 2015 and 2016. Two premises in Capitale-Nationale and two premises in Monteregie were very close, and therefore are overlapping on the map. Mapping was performed in ArcGIS version 10.6, using reference maps from Statistics Canada (2016 census).

Among the sampled horses, 48% were geldings, 49% were female and 3% were stallions. The mean age was 12 years old with a range from 2 to 30 years old.

3.1. Indicator Collection

E. coli isolates were detected in 209 (93%) of the 225 rectal swabs. A total of 609 *E. coli* isolates were selected from 209 samples, originating from the 32 premises.

The prevalence estimates of horses shedding non-susceptible isolates per antimicrobial and of the premises housing those horses are shown in Figure 2. Over 40% of horses shed isolates that were non-susceptible to ampicillin, streptomycin or amoxicillin + clavulanic acid. Over 60% of premises housed horses that shed isolates non-susceptible to streptomycin, nalidixic acid, folate pathway inhibitors (trimethoprim-sulfamethoxazole and sulfisoxazole), ampicillin, amoxicillin + clavulanic acid or tetracycline.

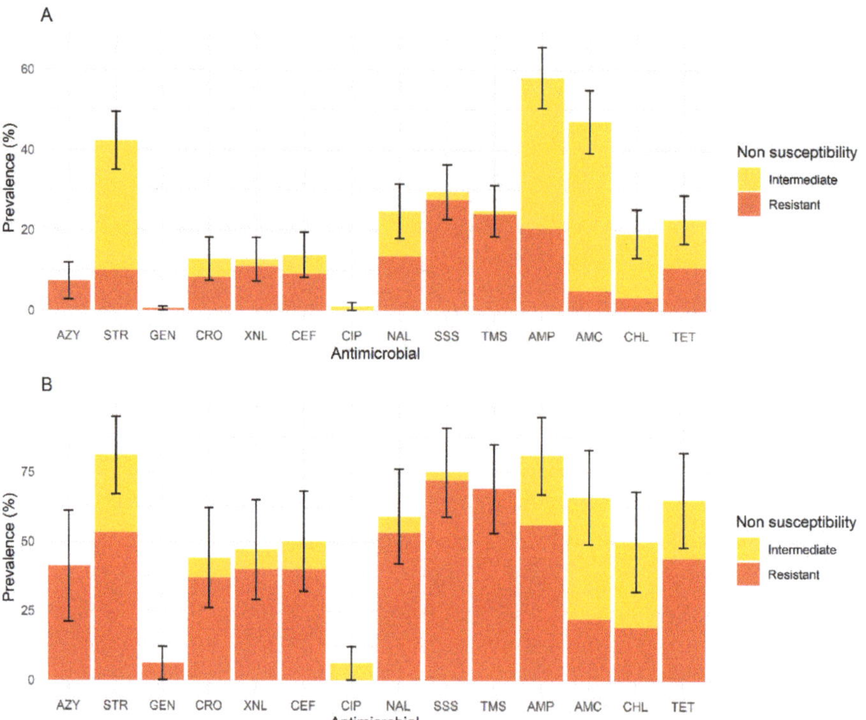

Figure 2. Prevalence estimates (%) of non-susceptibility (yellow and red) for each antimicrobial, at the horse level (**A**), and at the premises level (**B**), in a cross-sectional study of 209 healthy adult horses, in 32 premises, performed in 2015 and 2016, in Quebec. Bars represent 95% confidence intervals for prevalence of non-susceptible isolates. The proportion of resistant isolates for each antimicrobial is presented in red. A total of 609 isolates were tested. Abbreviations: AZY = azithromycin, STR = streptomycin, GEN = gentamicin, CRO = ceftriaxone, XNL = ceftiofur, CEF = cefoxitin, CIP = ciprofloxacin, NAL = nalidixic acid, SSS = sulfisoxazole, TMS = trimethoprim–sulfamethoxazole, AMP = ampicillin, AMC = amoxicillin/clavulanic acid, CHL = chloramphenicol, TET = tetracycline.

As illustrated in Figure 2, in this collection, non-susceptibility to third generation cephalosporins (ceftiofur and ceftriaxone), was observed in 12.8% of horses and 46.8% of premises. We did not identify any *bla* genes as tested by PCR in these isolates.

The prevalence estimate of non-susceptibility to nalidixic acid, a first-generation quinolone, was high (24.7% of horses and 59.4% of premises). In contrast, the estimated prevalence of non-susceptibility to ciprofloxacin, a fluoroquinolone, was relatively low (1.0% of horses and 6.3% of premises).

Prevalence estimates (%) of horses shedding isolates non-susceptible to ≥ 1, 3, 5, 7 or 9 classes of antimicrobials and premises housing these horses, are summarized in Table 1. The prevalence of horses shedding isolates non-susceptible to at least one antimicrobial and MDR isolates were high (80.0% and 46.3%, respectively). Of the 32 premises, 81.3% housed at least one horse shedding MDR isolates. In addition, 1.4% of horses shed isolates non-susceptible to nine classes of antimicrobials, and therefore had a potential for extensive resistance [26].

Table 1. Prevalence estimates (%) with 95% confidence intervals (95% CI) of healthy adult horses shedding *E. coli* isolates that are non-susceptible to more than 1, 3, 5, 7 or 9 classes of antimicrobials and premises housing these horses based on the indicator collection results in a cross-sectional study of 209 horses in 32 premises in Quebec in 2015 and 2016. Abbreviations: CI = confidence interval, MDR = multidrug-resistant.

	Indicator Collection			
Number of Resistant Antimicrobial Classes	**Horse Level (*n* = 209)**		**Premises Level (*n* = 32)**	
	%	95% CI	%	95% CI
≥1	80.0	69.8–90.2	96.9	90.5–100
≥3 (MDR)	46.3	34.5–58.0	81.3	67.0–95.5
≥5	15.3	8.5–22.3	53.1	34.8–71.4
≥7	3.9	1.5–6.2	25.0	9.1–40.9
≥9	1.4	0–3.2	9.4	0–20.1

3.2. ESBL/AmpC Collection

A total of 7.3% [95% CI 0–17.6] of the 209 horses shed ESBL/AmpC isolates non-susceptible to ceftriaxone, therefore belonging to the ESBL/AmpC collection, and 18.8% [95% CI 4.5–33] of the 32 premises housed these horses.

Non-susceptibility pattern of the 74 isolates of this collection originating from the 17 positive horses, found in six premises, is shown in Figure 3. All isolates were non-susceptible to ampicillin and ceftriaxone, although three isolates presented susceptibility to ceftiofur and 60 isolates presented susceptibility to cefoxitin, a cephamycin, also considered as a second-generation cephalosporin [28].

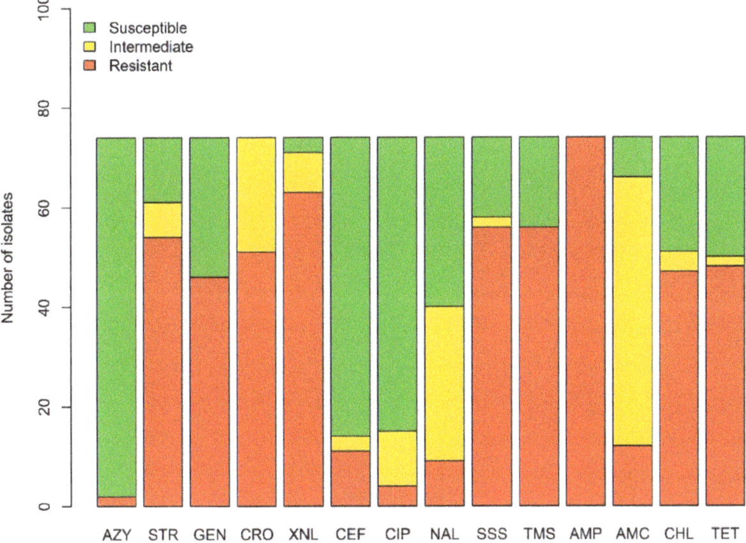

Figure 3. Susceptibility profiles of *E. coli* isolates in the ESBL/AmpC collection, in a cross-sectional study performed on healthy adult horses, in Quebec in 2015 and 2016 (n = 74 isolates distributed in 17 horses among 6 premises). Abbreviations: AZY = azithromycin, STR = streptomycin, GEN = gentamicin, CRO = ceftriaxone, XNL = ceftiofur, CEF = cefoxitin, CIP = ciprofloxacin, NAL = nalidixic acid, SSS = sulfisoxazole, TMS = trimethoprim–sulfamethoxazole, AMP = ampicillin, AMC = amoxicillin/clavulanic acid, CHL = chloramphenicol, TET = tetracycline.

Non-susceptibility to aminoglycosides (gentamicin and streptomycin), tetracycline, folate inhibitors (trimethoprim-sulfonamides, sulfizoxasole) and chloramphenicol were present in over 60% of the isolates.

A total of 54.1% of isolates were non-susceptible to a first-generation quinolone (nalidixic acid) and 20.3% of isolates were non-susceptible to ciprofloxacin, a fluoroquinolone, in this collection. These isolates were therefore non-susceptible to two families of antimicrobial classified as having the highest priority in human medicine by both Canadian Health and the WHO [21,22].

The main ESBL genes identified were $bla_{CTX\text{-}M\text{-}1}$ (43/74 tested isolates) and bla_{SHV} (15/74), four isolates carried a combination of $bla_{CTX\text{-}M\text{-}1}$ and bla_{SHV}. Nine isolates carried the AmpC gene $bla_{CMY\text{-}2}$. In four isolates we could not detect tested ESBL/AmpC genes by PCR.

3.3. Risk Factors

A total of 13 potential risk factors were derived from the questionnaire (Table 2). Eleven were considered at the individual level and two were considered at the premise level.

Table 2. Descriptive statistics and *p*-value (Wald test) from univariable multilevel logistic regression analyses of potential risk factors for MDR status in horses in a cross-sectional study performed on healthy adult horses, in Quebec, in 2015 and 2016. In bold are the factors that were retained for the multivariable analysis.

Putative Risk Factors	Number of Horses	% of MDR-Positive Horses	*p*-Value
Horse-level			
Transportation out of the horse's premises within the last 3 months			**0.11**
Yes	31	41.9	
No	69	29.0	
Participation to an equestrian event within the last 3 months			**0.13**
Yes	19	57.9	
No	92	29.3	
Housing			0.28
Stable	46	39.1	
Pasture [1]	64	29.7	
Activity			0.73
Sport (competition)	39	43.6	
Leisure	66	30.3	
Reproduction	4	25.0	
The horse presents a chronic disease			0.47
Yes	18	22.2	
No	123	39.0	
The horse presented an infection (diagnosed by the veterinarian) within the last 3 months			**0.03**
Yes	12	75.0	
No	124	34.7	
The horse presented diarrhea within the last 3 months			0.91
Yes	5	20.0	
No	61	23.0	
The horse was hospitalized within the last 3 months			The model did not converge
Yes	2	100	
No	137	37.2	

Table 2. *Cont.*

Putative Risk Factors	Number of Horses	% of MDR-Positive Horses	*p*-Value
The horse has undergone surgery within the last 3 months			The model did not converge
- Yes	1	100	
- No	137	38.0	
The horse has been medically treated within the last three months (all treatment considered)			**0.02**
- Yes	39	53.9	
- No	94	30.9	
The horse presented with colic within the last 3 months			0.81
- Yes	5	40	
- No	131	37.4	
Premise-level			
Total number of horses in the premises [2]			0.26
- Less than 15	101	45.5	
- 15 and more	109	48.6	
Number of staff taking care of horses daily [3]			**0.01**
- Less than 5 persons	68	23.5	
- 5 persons and more	38	47.4	

[1] Defined as a horse that stays on pasture at night and has a shelter in the pasture. [2] Categorization was done *a posteriori*, based on the mean of the number of horses in the premises we sampled. [3] This variable was already categorized in the questionnaire.

Data with missing values, representing almost half of the dataset, were excluded from modeling.

A total of five risk factors were selected for multivariate modeling (all $p < 0.20$ in univariable logistic regressions) for the MDR outcome. The variable "The horse presented an infection" was then excluded as it was associated with "The horse has been medically treated within the last 3 months". The variable "Transportation within the last 3 months" was excluded because it was associated with "Participating in an equestrian event within the last 3 months".

According to the multivariate model, the odds of being an MDR horse were 3.5 times higher ($p = 0.03$) among the horses that had participated to an equestrian event within the last three months and 3.4 times higher ($p = 0.01$) if the horse was in a premise where the staff were composed of more than five persons (Table 3). Visual assessment of residuals at the premise level suggested that our model fitted the data.

Table 3. Parameter estimates and odds ratios from a multivariable regression modeling MDR positive status at the horse level, based on the results of a cross-sectional study performed on 32 premises and 209 healthy adult horses, sampled in Quebec, in summers 2015 and 2016. The estimated variance at the premises level was 0.171 (standard error of 0.316).

Risk Factor for the Outcome	Odds Ratios		
	Estimate	95% CI	*p*
The horse has participated in an equestrian event within the last three months (yes vs. no)	3.5	1.1, 11.1	0.03
Number of staff taking care of horses daily (5 persons and more vs. less than 5 persons)	3.4	1.3, 8.7	0.01

Interaction between the two variables of the final multivariable model were checked but were not significant ($p = 0.41$, Wald test) and thus not kept in the model.

For the ESBL/AmpC outcome, considering the high percentage of missing data and low number of positive horses, no statistical modelling was performed.

4. Discussion

The present study illustrates that the fecal microbiota of healthy horses in Quebec, Canada, harbor MDR and ESBL/AmpC *E. coli* isolates. The prevalence of horses shedding ESBL/AmpC *E. coli* isolates (7.3%) is comparable to that which was detected phenotypically in the United Kingdom in 2012 (6.3%) [6]. Nevertheless, at the premise level, it seems that the prevalence in Quebec (18.8%) is inferior to the prevalence reported in France (29.0%) [7]. However, these regional differences in apparent prevalence might be related to a higher sensitivity in the detection of positive premises in the study in France, considering that in France we tested more horses per premise (between six and 36 horses per premises) and ESBL/AmpC isolates were detected by two enrichment methods. The prevalence we found in horses in Quebec contrasts with the 1% prevalence observed in Ontario among 188 healthy dogs in 2009 [29], although, even if the calculation methods are not the same, this is still lower than the 26.5% of fecal carriage of ESC-resistant Enterobacteriaceae in healthy dogs in Ontario in 2018 [30]. The prevalence of horses shedding ESBL/AmpC *E. coli* isolates in Quebec is higher than that reported in Sable Island horses, where 1/508 horses shedding an ESBL gene [31] ($bla_{CTX-M-1}$) was found. This is not surprising because our horse population is in contact with the populations of other species in which ESBL/AmpC genes have been detected, such as pigs [32], poultry [33], cattle [34] and humans [35], underlining the importance of the one health approach [1,36] to address the problem.

Our study reported the presence of isolates that are non-susceptible to nine classes of antimicrobial in an indicator collection of *E. coli* from horses for the first time to our knowledge, which is worrisome. Although these isolates may be commensals, it is possible that putative resistance genes are carried by mobile genetic elements, such as plasmids, and are therefore transmissible to potential pathogenic or zoonotic strains. The dissemination of extensive resistance to pathogenic strains could lead to an increased risk of complications in the treatment of infections caused by these strains.

Enrofloxacin, a fluoroquinolone, is classified as having a very high importance in human medicine [22] and is approved for veterinary use in equine medicine. Resistance to quinolones is known to be acquired and is mostly due to the apparition of chromosomal mutations, although resistance genes carried by plasmids have also been reported [37]. Often, the chromosomal mutations appear consecutively and are localized on the genes *gyrA* and *parC* (coding for gyrase and topoisomerase, respectively, both involved in the DNA synthesis). The number of mutations is proportionate to the minimal inhibitory concentration (the more the higher). Hence, non-susceptibility to nalidixic acid is generally precursory for fluoroquinolone treatment failure [38]. In the indicator collection, we detected 24.7% of horses and 59.4% of premises presenting a non-susceptibility to nalidixic acid, suggesting that enrofloxacin should be used with caution, to maintain its efficacy in horses in Quebec. In the ESBL/AmpC collection, we found 14 isolates presenting a non-susceptibility to both 3rd generation cephalosporin and fluoroquinolones. Even though these isolates are unlikely to be pathogenic, they still represent a risk of dissemination due to their high capacity to resist antimicrobial pressure. They could acquire virulence genes through the transfer of plasmids thus becoming a threat for public and/or equine health.

In our study, the predominant ESBL gene found was $bla_{CTX-M-1}$. ESBL of the CTX-M family have become a public health concern in the last two decades, their incidence and diversity having increased dramatically during this time and have overridden other ESBL variants such as bla_{TEM} and bla_{SHV} in gram negative bacteria [15]. The bla_{CTX-M} encoded ESBL family is characterized by the ability to inhibit 3rd and 4th generation cephalosporins and monobactams, but not cephamycins and carbapenems. These ESBLs are also known to be susceptible to β-lactam inhibitors. However, no cephamycin or penicillin/β-lactam inhibitor combinations are approved or used off-label (to the authors' knowledge) to treat horses. The predominance of $bla_{CTX-M-1}$ suggests a global dissemination of this gene in the equine population both in Europe and in North America. The absence of other variants of bla_{CTX-M} in the Quebec horse population contrasts with the high diversity of bla_{CTX-M} found in the healthy equine population in France and throughout Europe [7]. This suggests that the presence of this family of genes may have occurred later in North America than in Europe, and that the genes may not yet have had the time to diversify.

We detected the AmpC resistance gene bla_{CMY-2} in several horses. This gene has been frequently found in poultry and pigs, including in Quebec [32,39]. Although this gene has previously been identified in one healthy horse in France, the fact that we identified it in several healthy horses in Quebec suggests the possibility of AmpC gene spread between animal species. Indeed, horses can be in contact with other animal species, including dogs, cats, poultry among others, in the premises.

Even though we detected 12.8% of horses carrying isolates non-susceptible to 3rd generation cephalosporins in the indicator collection, none of these isolates carried the tested ESBL/AmpC genes, similar to what had been found in the indicator collection of our previous work [7]. These findings suggest that other mechanisms of resistance to cephalosporins (for example, alteration of the protein binding protein) may be present in the population. These alternative mechanisms are less likely to spread through plasmids but could impact cephalosporin efficacy, and therefore could affect equine welfare. We also found four isolates of the ESBL/AmpC collection in which we could not identify a *bla* gene. This could indicate that other, less common, *bla* genes are present in the horse population.

Among the risk factors model selected for modeling, the correlation between the variable "The horse presented an infection" and "The horse has been medically treated within the last 3 months" was to be expected, because a horse with an infection is often treated for this infection. The medical treatment of the horse was considered more biologically relevant to influence the shedding of MDR *E. coli* rather than the infection itself. However, this variable was not retained in the final model, perhaps because of the absence of specific information about the type of treatment, which could include treatments other than antimicrobials.

A correlation between "Transportation within the last 3 months" and "Participating in an equestrian event within the last 3 months" was also observed, which was not surprising, as horses which participate in an equestrian event are often transported to the equestrian event. We chose to consider participation in the equestrian event because of the possibility of transmission of antimicrobial resistance genes inter- and intraspecies during the event.

To our knowledge, we demonstrated for the first time that participation in an equestrian event was a risk factor for shedding MDR isolates at the horse level. Considering the correlation between the horse participation in an equestrian event and transportation, this effect could also be driven by contacts occurring during transportation. Based on this association, we could suggest isolating horses that are participating in equestrian events or at least the implementation of appropriate biosecurity measures. As an example, limiting contact between these horses and horses that stay at home or handling horses that stayed at home before horses that travelled might be beneficial to limit antimicrobial gene dissemination. However, more longitudinal studies are needed to establish the duration of shedding, and therefore be more accurate in these recommendations.

Our results suggest that a higher number of persons taking care of horses daily increases the risk of detecting MDR isolates in the horse's intestinal microbiota. We previously documented that this factor was associated with a higher risk of detecting ESBL/AmpC isolates in the healthy equine population in France [7]. The fact that this variable was found to be significant in both studies is noteworthy. Indeed, such information is easily obtained, and therefore could be helpful for elaborating guidelines to improve equine health. It could help equine veterinarians in defining "at-risk" equine populations and encourage the use of antimicrobial susceptibility testing in these populations.

The absence of a probability sampling method in our study might affect the representativeness of our prevalence estimate. The extrapolation of such estimate to the general equine population should be made cautiously, as the horses selected for our study are more likely representative of a subpopulation of horses under regular veterinary follow-up examination. Another limitation of our study is the recruitment of a smaller sample size of horses than planned, combined with a high percentage of missing data for the questionnaire among recruited horses (almost 50%), thus reducing the precision of the prevalence estimates and statistical power of the risk factor analyses. The low participation rate could be due to a lack of awareness of the importance of antimicrobial resistance in the equine industry. A higher proportion of missing values were present in horses shedding MDR isolates. This could be due to some regional differences and/or owner characteristics influencing both the risk of MDR and interest to participate in our study. The validity of our results depends on the absence of association between response rate and exposure to identified risk factors. Such association seems unlikely considering that the MDR status and associated risk factors were unknown for both horse owners and veterinarians at the time of data collection.

A valuable follow up to this study would be to sample the veterinarians and owners of these horses and see if there is a correlation between horses and horse handlers for the carriage of ESBL/AmpC producing *E. coli*. Another interesting follow up would be to repeat the study a few years after the regulations (see introduction) have been set up and see if these have made a difference.

5. Conclusions

In conclusion, we found a noteworthy prevalence of ESBL/AmpC genes and MDR isolates in the fecal microbiota of healthy horses in Quebec. Surveillance of ESBL/AmpC gene dissemination and the quantification of MDR isolates would be beneficial to characterize the nature and the extent of the risk they represent, with the aim of limiting their transmission between horses, but also to other species including humans and to the environment. The detection of risk factors for MDR shedding could be used to help equine veterinarians in managing at-risk populations.

Supplementary Materials: The following are available online at http://www.mdpi.com/2076-2615/10/3/523/s1, Figure S1: Questionnaire in French, Figure S2: Questionnaire in English.

Author Contributions: Conceptualization, M.d.L., J.A., J.M.F.; methodology, M.d.L., J.A., J.M.F.; software, M.d.L., J.A.; validation, J.A.; formal analysis, M.d.L., J.A.; investigation, M.d.L.; resources, M.d.L., J.M.F.; data curation, M.d.L., J.A., J.M.F.; writing—original draft preparation, M.d.L.; writing—review and editing, J.A., J.M.F.; visualization, M.d.L., J.A., J.M.F.; supervision, J.A., J.M.F.; project administration, M.d.L.; funding acquisition, M.d.L., J.M.F. All authors have read and agreed to the published version of the manuscript.

Funding: This research was funded by the Ecl Laboratory, the AVEQ and the MAPAQ.

Acknowledgments: We thank all the equine veterinarians from Quebec that participated to our study.

Conflicts of Interest: The authors declare no conflict of interest.

References

1. World Health Organization. *Antimicrobial Resistance: Global Report on Surveillance*; World Health Organization: Geneva, Sweden, 2014.

2. Hariharan, H.; Barnum, D.A.; Mitchell, W.R. Drug resistance among pathogenic bacteria from animals in Ontario. *Can. J. Comp. Med.* **1974**, *38*, 213–221.

3. van Spijk, J.; Schmitt, S.; Schoster, A. Infections caused by multidrug-resistant bacteria in an equine hospital (2012–2015). *Equine Vet. Educ.* **2019**, *31*, 653–658. [CrossRef]

4. Jokisalo, J.; Bryan, J.; Legget, B.; Abbott, Y.; Katz, L. Multiple-drug resistant Acinetobacter baumannii bronchopneumonia in a colt following intensive care treatment. *Equine Vet. Educ.* **2010**, *22*, 281–286. [CrossRef]

5. Giguere, S.; Berghaus, L.J.; Willingham-Lane, J.M. Antimicrobial resistance in *Rhodococcus equi*. *Microbiol. Spectr.* **2017**, *5*. [CrossRef]

6. Maddox, T.W.; Clegg, P.D.; Diggle, P.J.; Wedley, A.L.; Dawson, S.; Pinchbeck, G.L.; Williams, N.J. Cross-sectional study of antimicrobial-resistant bacteria in horses. Part 1: Prevalence of antimicrobial-resistant *Escherichia coli* and methicillin-resistant *Staphylococcus aureus*. *Equine Vet. J.* **2012**, *44*, 289–296. [CrossRef] [PubMed]

7. de Lagarde, M.; Larrieu, C.; Praud, K.; Schouler, C.; Doublet, B.; Salle, G.; Fairbrother, J.M.; Arsenault, J. Prevalence, risk factors, and characterization of multidrug resistant and extended spectrum beta-lactamase/AmpC beta-lactamase producing *Escherichia coli* in healthy horses in France in 2015. *J. Vet. Intern. Med.* **2019**, *33*, 902–911. [CrossRef] [PubMed]

8. Aarestrup, F.M.; Hasman, H.; Veldman, K.; Mevius, D. Evaluation of eight different cephalosporins for detection of cephalosporin resistance in *Salmonella enterica* and *Escherichia coli*. *Microb. Drug Resist.* **2010**, *16*, 253–261. [CrossRef]

9. Maddox, T.W.; Clegg, P.D.; Williams, N.J.; Pinchbeck, G.L. Antimicrobial resistance in bacteria from horses: Epidemiology of antimicrobial resistance. *Equine Vet. J.* **2015**, *47*, 756–765. [CrossRef]

10. DebRoy, C.; Roberts, E.; Jayarao, B.M.; Brooks, J.W. Bronchopneumonia associated with extraintestinal pathogenic *Escherichia coli* in a horse. *J. Vet. Diagn. Investig.* **2008**, *20*, 661–664. [CrossRef]

11. Government of Canada. *Canadian Antimicrobial Resistance Surveillance Systeme—Report*; Public Health Agency: Ottawa, ON, Canada, 2016. Available online: http://publications.gc.ca/pub?id=9.512584&sl=1 (accessed on 1 September 2016).

12. Rubin, J.E.; Pitout, J.D. Extended-spectrum beta-lactamase, carbapenemase and AmpC producing *Enterobacteriaceae* in companion animals. *Vet. Microbiol.* **2014**, *170*, 10–18. [CrossRef]

13. Johns, I.C.; Adams, E.L. Trends in antimicrobial resistance in equine bacterial isolates: 1999–2012. *Vet. Rec.* **2015**, *176*, 334. [CrossRef] [PubMed]

14. Ewers, C.; Bethe, A.; Semmler, T.; Guenther, S.; Wieler, L.H. Extended-spectrum beta-lactamase-producing and AmpC-producing *Escherichia coli* from livestock and companion animals, and their putative impact on public health: A global perspective. *Clin. Microbiol. Infect.* **2012**, *18*, 646–655. [CrossRef] [PubMed]

15. Cantón, R.; González-Alba, J.; Galán, J. CTX-M enzymes: Origin and diffusion. *Front. Microbiol.* **2012**, *3*, 110. [CrossRef] [PubMed]

16. Carattoli, A. Plasmids in Gram negatives: Molecular typing of resistance plasmids. *Int. J. Med. Microbiol.* **2011**, *301*, 654–658. [CrossRef]

17. Mathers, A.J.; Peirano, G.; Pitout, J.D. The role of epidemic resistance plasmids and international high-risk clones in the spread of multidrug-resistant *Enterobacteriaceae*. *Clin. Microbiol. Rev.* **2015**, *28*, 565–591. [CrossRef]

18. Roer, L.; Overballe-Petersen, S.; Hansen, F.; Schonning, K.; Wang, M.; Roder, B.L.; Hansen, D.S.; Justesen, U.S.; Andersen, L.P.; Fulgsang-Damgaard, D.; et al. *Escherichia coli* sequence type 410 is causing new international high-risk clones. *mSphere* **2018**, *3*, e00337-18. [CrossRef]

19. Huijbers, P.M.; de Kraker, M.; Graat, E.A.; van Hoek, A.H.; van Santen, M.G.; de Jong, M.C.; van Duijkeren, E.; de Greeff, S.C. Prevalence of extended-spectrum beta-lactamase-producing *Enterobacteriaceae* in humans living in municipalities with high and low broiler density. *Clin. Microbiol. Infect.* **2013**, *19*, E256–E259. [CrossRef]

20. Barbier, F.; Pommier, C.; Essaied, W.; Garrouste-Orgeas, M.; Schwebel, C.; Ruckly, S.; Dumenil, A.-S.; Lemiale, V.; Mourvillier, B.; Clec'h, C.; et al. Colonization and infection with extended-spectrum β-lactamase-producing Enterobacteriaceae in ICU patients: What impact on outcomes and carbapenem exposure? *J. Antimicrob. Chemother.* **2016**, *71*, 1088–1097. [CrossRef]

21. Health Canada. Categorization of Antimicrobial Drugs Based on Importance in Human Medicine. 2019. Available online: https://www.canada.ca/en/health-canada/services/drugs-health-products/veterinary-drugs/antimicrobial-resistance/categorization-antimicrobial-drugs-based-importance-human-medicine.html (accessed on 1 April 2009).

22. World Health Organization. *Critically Important Antimicrobials for Human medicine: Ranking of Antimicrobial Agents for Risk Management of Antimicrobial Resistance due to Non-Human Use*; World Health Organization: Geneva, Sweden, 2017.

23. Gouvernment of Quebec. Decret 1110–2018. Loi sur la protection des animaux (chapitre P-42). In *Administration de Certains Médicaments-Modification*; Editeur Officiel du Quebec: Quebec, QC, Canada, 2019.

24. Maddox, T.W.; Pinchbeck, G.L.; Clegg, P.D.; Wedley, A.L.; Dawson, S.; Williams, N.J. Cross-sectional study of antimicrobial-resistant bacteria in horses. Part 2: Risk factors for faecal carriage of antimicrobial-resistant *Escherichia coli* in horses. *Equine Vet. J.* **2012**, *44*, 297–303. [CrossRef]

25. Martins, M.T.; Rivera, I.G.; Clark, D.L.; Stewart, M.H.; Wolfe, R.L.; Olson, B.H. Distribution of *uidA* gene sequences in *Escherichia coli* isolates in water sources and comparison with the expression of beta-glucuronidase activity in 4-methylumbelliferyl-beta-D-glucuronide media. *Appl. Environ. Microbiol.* **1993**, *59*, 2271–2276. [CrossRef]

26. Magiorakos, A.P.; Srinivasan, A.; Carey, R.B.; Carmeli, Y.; Falagas, M.E.; Giske, C.G.; Harbarth, S.; Hindler, J.F.; Kahlmeter, G.; Olsson-Liljequist, B.; et al. Multidrug-resistant, extensively drug-resistant and pandrug-resistant bacteria: An international expert proposal for interim standard definitions for acquired resistance. *Clin. Microbiol. Infect.* **2012**, *18*, 268–281. [CrossRef] [PubMed]

27. Agerso, Y.; Aarestrup, F.M.; Pedersen, K.; Seyfarth, A.M.; Struve, T.; Hasman, H. Prevalence of extended-spectrum cephalosporinase (ESC)-producing *Escherichia coli* in Danish slaughter pigs and retail meat identified by selective enrichment and association with cephalosporin usage. *J. Antimicrob. Chemother.* **2012**, *67*, 582–588. [CrossRef] [PubMed]

28. Giguère, S.; Prescott, J.F.; Dowling, P.M. *Antimicrobial Therapy in Veterinary Medicine*; John Wiley & Sons: New York, NY, USA, 2013.

29. Murphy, C.; Reid-Smith, R.J.; Prescott, J.F.; Bonnett, B.N.; Poppe, C.; Boerlin, P.; Weese, J.S.; Janecko, N.; McEwen, S.A. Occurrence of antimicrobial resistant bacteria in healthy dogs and cats presented to private veterinary hospitals in southern Ontario: A preliminary study. *Can. Vet. J.* **2009**, *50*, 1047–1053. [PubMed]

30. Zhang, P.L.C.; Shen, X.; Chalmers, G.; Reid-Smith, R.J.; Slavic, D.; Dick, H.; Boerlin, P. Prevalence and mechanisms of extended-spectrum cephalosporin resistance in clinical and fecal Enterobacteriaceae isolates from dogs in Ontario, Canada. *Vet. Microbiol.* **2018**, *213*, 82–88. [CrossRef] [PubMed]

31. Timonin, M.E.; Poissant, J.; McLoughlin, P.D.; Hedlin, C.E.; Rubin, J.E. A survey of the antimicrobial susceptibility of *Escherichia coli* isolated from Sable Island horses. *Can. J. Microbiol.* **2017**, *63*, 246–251. [CrossRef]

32. Jahanbakhsh, S.; Letellier, A.; Fairbrother, J.M. Circulating of CMY-2 beta-lactamase gene in weaned pigs and their environment in a commercial farm and the effect of feed supplementation with a clay mineral. *J. Appl. Microbiol.* **2016**, *121*, 136–148. [CrossRef]

33. Verrette, L.; Fairbrother, J.M.; Boulianne, M. Effect of cessation of ceftiofur and substitution with lincomycin-spectinomycin on extended-spectrum-beta-lactamase/AmpC genes and multidrug resistance in *Escherichia coli* from a Canadian broiler production pyramid. *Appl. Environ. Microbiol.* **2019**, *85*, e00037-19. [CrossRef]

34. Allen, K.J.; Poppe, C. Occurrence and characterization of resistance to extended-spectrum cephalosporins mediated by beta-lactamase CMY-2 in Salmonella isolated from food-producing animals in Canada. *Can. J. Vet. Res.* **2002**, *66*, 137–144.

35. Peirano, G.; Costello, M.; Pitout, J.D. Molecular characteristics of extended-spectrum beta-lactamase-producing *Escherichia coli* from the Chicago area: high prevalence of ST131 producing CTX-M-15 in community hospitals. *Int. J. Antimicrob. Agents* **2010**, *36*, 19–23. [CrossRef]

36. WHO (World Health Organisation). One Health. 2017. Available online: https://www.who.int/features/qa/one-health/en/ (accessed on 1 February 2017).

37. Jacoby, G.A. Mechanisms of resistance to quinolones. *Clin. Infect. Dis.* **2005**, *41* (Suppl. 2), S120–S126. [CrossRef]

38. Robert, J.; Cambau, E.; Grenet, K.; Trystram, D.; Pean, Y.; Fievet, M.H.; Jarlier, V. Trends in quinolone susceptibility of *Enterobacteriaceae* among inpatients of a large university hospital: 1992–1998. *Clin. Microbiol. Infect.* **2001**, *7*, 553–561. [CrossRef] [PubMed]

39. Aslam, M.; Checkley, S.; Avery, B.; Chalmers, G.; Bohaychuk, V.; Gensler, G.; Reid-Smith, R.; Boerlin, P. Phenotypic and genetic characterization of antimicrobial resistance in Salmonella serovars isolated from retail meats in Alberta, Canada. *Food Microbiol.* **2012**, *32*, 110–117. [CrossRef] [PubMed]

Article

Multidrug-Resistant ESBL/AmpC-Producing *Klebsiella pneumoniae* Isolated from Healthy Thoroughbred Racehorses in Japan

Eddy Sukmawinata [1], Ryoko Uemura [2,3,*], Wataru Sato [2], Myo Thu Htun [2] and Masuo Sueyoshi [1,2,3]

[1] Graduate School of Medicine and Veterinary Medicine, University of Miyazaki, Miyazaki 889-1692, Japan; eddyswinata@gmail.com (E.S.); a0d802u@cc.miyazaki-u.ac.jp (M.S.)

[2] Department of Veterinary Sciences, Faculty of Agriculture, University of Miyazaki, Miyazaki 889-2192, Japan; wataru9356@gmail.com (W.S.); drmyothu.htun@gmail.com (M.T.H.)

[3] Center for Animal Diseases Control, University of Miyazaki, Miyazaki 889-2192, Japan

* Correspondence: uemurary@cc.miyazaki-u.ac.jp; Tel.: +81-985-58-7283

Received: 28 January 2020; Accepted: 20 February 2020; Published: 25 February 2020

Simple Summary: Extended-spectrum β-lactamases (ESBLs) and AmpC β-lactamases (AmpCs) have been recognized as an emerging global problem in humans and animals. These enzymes provide a mechanism of resistance by inactivating β-lactam antibiotics and are mostly encoded on plasmids, which can be easily transmitted to other bacteria in humans, animals, and the environment. Several clinical diseases caused by *Klebsiella* spp. infection have been confirmed in the horse community. The emergence of antimicrobial resistance in *Klebsiella* spp. increases the risk of treatment failure in infected horses. In this study, we investigated the presence of ESBL/AmpC-producing *Klebsiella* spp. isolated from healthy Thoroughbred racehorses in Japan. The results showed that ESBL/AmpC-producing *Klebsiella pneumoniae* (ESBL/AmpC-KP) isolated from horses have co-resistance to other β-lactam antibiotics as multidrug-resistant (MDR) bacteria. Genetic relatedness analysis suggested that plasmid-mediated AmpC-KP clones may spread between horses. This is the first study to show *K. pneumoniae* carrying MDR plasmid-mediated AmpC isolated from racehorses. Continuous monitoring antimicrobial resistance to this species is required in order to control the spread of MDR ESBL/AmpC-KP in the racehorse community.

Abstract: Extended-spectrum β-lactamase (ESBL)- and AmpC β-lactamase (AmpC)-producing *Klebsiella* spp. have become a major health problem, leading to treatment failure in humans and animals. This study aimed to evaluate the presence of ESBL/AmpC-producing *Klebsiella* spp. isolated from racehorses in Japan. Feces samples from 212 healthy Thoroughbred racehorses were collected from the Japan Racing Association Training Centers between March 2017 and August 2018. ESBL/AmpC-producing *Klebsiella* spp. were isolated using selective medium containing 1 μg/mL cefotaxime. All isolates were subjected to bacterial species identification (MALDI-TOF MS), antimicrobial susceptibility test (disk diffusion test), characterization of resistance genes (PCR), conjugation assay, and genetic relatedness (multilocus sequence typing/MLST). Twelve ESBL/AmpC-producing *Klebsiella pneumoniae* (ESBL/AmpC-KP) were isolated from 3.3% of horse samples. Antimicrobial resistance profiling for 17 antimicrobials showed all ESBL/AmpC-KP were multidrug-resistant (MDR). Only 1 isolate was confirmed as an ESBL producer (*bla*CTX-M-2-positive), whereas the other 11 isolates were plasmid-mediated AmpC (pAmpC) producers (*bla*CMY positive). On the basis of MLST analysis, the ESBL-KP isolate was identified as sequence type (ST)-133 and four different STs among AmpC-KP isolates, ST-145, ST-4830, ST-4831, and ST-4832, were found to share six of the seven loci constituting a single-locus variant. This is the first study to show *K. pneumoniae* carrying MDR pAmpC isolated from a racehorse.

Keywords: extended-spectrum β-lactamase; AmpC β-lactamase; *Klebsiella pneumoniae*; horse; multidrug resistance

1. Introduction

Klebsiella spp. is a normal intestinal bacteria in horses [1] and is ubiquitous in the environment [2]. However, some studies have reported *Klebsiella* spp. as a causal agent for infections in horses, such as mares with metritis and cervicitis, foals with septicemia and pneumonia [3], and pneumonia in adult horses [2], and disease severity depends on the pathogenicity of the strains [1]. In the horse industry, about 25%–60% of economic losses are caused by endometritis, and *Klebsiella pneumoniae* was reported as one of the causal infections that can be transmitted through the venereal route [4,5]. First-, second-, and third-generation cephalosporin has been used for treatment of bacterial infection in equine medicine for several years. Ceftiofur, which belongs to third-generation cephalosporin, is approved for used in horses and effective in treatment of *Klebsiella* infection. In special cases, such as septicemia in foals and respiratory tract disease in horses, cefquinome, which is a fourth-generation cephalosporin, is accepted for use in the United Kingdom [6]. However, the occurrence of antimicrobial resistance (AMR) in *Klebsiella* spp. has increased the risk of treatment failure [7].

Extended-spectrum β-lactamases (ESBLs) and AmpC β-lactamases (AmpCs) have emerged globally in humans and animals [8]. These enzymes can hydrolyze extended-spectrum cephalosporin [8], whereas AmpCs have a broader resistance spectrum to cephalosporins, including cephamycins (cefoxitin and cefotetan) [9]. ESBL and AmpC genes are mainly located on mobile genetic elements such as plasmids, which can be transferred to other bacteria in humans, animals, or the environment [10]. Nonetheless, AmpC is less frequently reported than ESBL [11,12]. β-Lactamase inhibitors such as clavulanic acid, sulbactam, and tazobactam have the effect of inhibiting the production of ESBL [9,13], but these have much less effect on AmpC β-lactamase [12].

Extended-spectrum β-lactamase-producing *Enterobacteriaceae* have gained special attention on AMR in horses due to their presence as a potentially zoonotic bacteria [14]. The CTX-M family of ESBL have been reported as the predominant type of ESBL after the TEM and SHV types [15], and more than 200 CTX-M variants have been identified worldwide [16]. On the other hand, some species of *Enterobacteriaceae* (such as *Enterobacter cloacae*, *Enterobacter aerogenes*, *Aeromonas* sp., *Citrobacter freundii*, *Providencia* sp., *Serratia marcescens*, *Hafnia alvei*, *Morganella morganii*, and *Pseudomonas aeruginosa*) have resistance to extended-spectrum cephalosporin, which may be caused by inducible chromosomal AmpC. Furthermore, plasmid-mediated AmpC (pAmpC) were identified from *Enterobacteriaceae* such as, *Klebsiella* spp., *Escherichia coli*, *Salmonella* spp., and *Proteus mirabilis* [17]. The distribution of pAmpC seems to be more frequent in animals than in humans [16]. Although ESBL/AmpC-producing *Klebsiella* spp. (ESBL/AmpC-K) are considered a major global concern, information is still lacking for AMR in horses [1]. Moreover, information on ESBL/AmpC-K in horses is unavailable in Japan. This study aimed to evaluate the presence of ESBL/AmpC-K isolated from healthy Thoroughbred racehorses in Japan. In addition, although carbapenems are rarely used in pet animals, these antimicrobials are frequently considered as the last option of treatment for ESBL/AmpC-producing bacteria infection [18,19]. In this work, all ESBL/AmpC-positive isolates were also tested for carbapenemase production.

2. Materials and Methods

2.1. Isolation of ESBL/AmpC-K

Feces samples from 212 healthy Thoroughbred racehorses were collected by veterinarians at the Japan Racing Association (JRA) between March 2017 and August 2018. Sampling locations were the Miho Training Center (103 samples) and Ritto Training Center (109 samples). No samples were from horses under treatment with antibiotics. Fresh feces samples from each individual horse were collected and stored in sterile plastic bags. Samples were sent immediately to our laboratory in a cooling box. ESBL/AmpC-K was screened on the basis of the European Committee on Antimicrobial Susceptibility Testing (EUCAST) guideline by using MacConkey agar (Nissui Pharmaceutical Co.,

Ltd., Tokyo, Japan) supplemented with 1 µg/mL cefotaxime (CTX; Duchefa Biochemie B.V. Haarlem, North Holland, the Netherlands) [20]. One to three colonies with pink, mucoid, and lactose fermented appearance were selected for species identification by using MALDI-TOF MS (Bruker, Billerica, MA, USA). All presumptive ESBL/AmpC-K isolates were stored frozen in trypticase soy broth (Nissui Pharmaceutical Co., Ltd., Tokyo, Japan) with 20% glycerol at −80 °C for further analysis. *Klebsiella pneumoniae* ATCC 700603 and *E. coli* ATCC 25922 were used as positive and negative control type strains, respectively.

All presumptive isolates were confirmed for ESBL and AmpC production by using the AmpC and ESβL Detection Set (D68C). All ESBL/AmpC positive isolates were further tested for carbapenemase production by Mastdiscs Combi Carba Plus (D73C), and the results were interpreted based on manufacturer guidelines (Mast Diagnostics, Merseyside, United Kingdom).

2.2. Antimicrobial Susceptibility Test

The antimicrobial susceptibility testing of all isolates were performed by disk diffusion assay to 17 antimicrobial agents belonging to 8 classes of antimicrobial, β-lactam (ampicillin 10 µg (ABPC), cefuroxime 30 µg (CXM), cefotaxime 30 µg (CTX), ceftazidime 30 µg (CAZ)), aminoglycoside (gentamicin 10 µg (GM), kanamycin 30 µg (KM), streptomycin 10 µg (SM), tetracycline (tetracycline 30 µg (TC), oxytetracycline 30 µg (OTC), doxycycline 30 µg (DOXY)), amphenicol (chloramphenicol 30 µg (CP)), polypeptide (colistin 10 µg (CL)), quinolone (nalidixic acid 30 µg (NA), norfloxacin 10 µg (NFLX), marbofloxacin 5 µg (MAR)), fosfomycin 200 µg (FOM), and folate antagonist-sulfonamide (trimethoprim/sulfamethoxazole 1.25/23.75 µg (STX)). Minimum inhibition zones were interpreted using the Clinical Laboratory Standard Institute (CLSI) criteria [21]. Multidrug-resistant (MDR) bacteria were termed to isolates that had resistance to at least three or more classes of antimicrobials [22]. *E. coli* ATCC 25922 strain was used for quality control.

2.3. Molecular Characterization of ESBL/AmpC-K

DNA from ESBL/AmpC-K isolates was extracted on the basis of the previously described method [23]. All ESBL/AmpC-positive isolates, the CTX-M-type β-lactamase and pAmpC genes were detected by multiplex PCR [24,25]. The bla_{TEM} and bla_{SHV} genes were identified by PCR and directly sequenced to confirm the type of β-lactamase [24]. Chromosomal AmpC, bla_{CMY}, *strA*, *strB*, *aphA1*, *tetA*, *tetB*, *cat*, and *floR* genes were identified by PCR [26–28], then one positive sample for each gene was selected for DNA sequencing to confirm the expected size, which was used as a positive control for other samples [24]. The results were analyzed with MEGA 7.0 (https://www.megasoftware.net/) and were examined with the National Center for Biotechnology Information, Basic Local Alignment Search Tool (NCBI BLAST) program (http://www.ncbi.nlm.nih.gov/blast/). The sequence types (STs) of *K. pneumoniae* were identified by multilocus sequence typing (MLST) on the basis of a previous report [29]. Novel STs were submitted to *Klebsiella pneumoniae* PubMLST and were termed as new STs (https://bigsdb.pasteur.fr/klebsiella/klebsiella.html).

2.4. Conjugation Assay

Transfer of antibiotic resistance was studied using conjugation for all ESBL/AmpC-K isolates. A plasmid-free and nalidixic acid-resistant (F⁻, Naʳ) of *E. coli* DH5α (Takara Bio Inc., Shiga, Japan) was used as a recipient strain, whereas all ESBL/AmpC-K resistant to NA served as donors. Conjugation was performed on the basis of our previous study [24].

2.5. Statistical Analysis

The antimicrobial susceptibility profile and the efficiency of conjugation were analyzed by descriptive statistics using Excel 2017 (version 15.40; Microsoft, Redmond, WA, USA).

3. Results

3.1. Resistance Phenotype

In this study, 12 ESBL/AmpC-producing *K. pneumoniae* (ESBL/AmpC-KP) were isolated from 7 (3.3%; 7/212) healthy Thoroughbred racehorse feces samples. Phenotypically, 11 isolates were confirmed as AmpC producers from 6 samples (2.8%; 6/212) that came from the Ritto Training Center, and only 1 sample from the Miho Training Center was confirmed as an ESBL producer. All ESBL/AmpC isolates were not identified as carbapenemase producers. All samples were resistant to ABPC, CXM, CTX, TC, OTC, DOXY, and FOM, followed by CAZ (83.3%; 10/12), GM (75.0%; 9/12), KM (66.7%; 8/12), SM (8.3%; 1/12), and CP (8.3%; 1/12). All isolates (100%; 12/12) were defined as MDR, meaning that they were resistant to at least three classes of antimicrobial. ESBL, AmpC, and resistance phenotype patterns are shown in Figure 1.

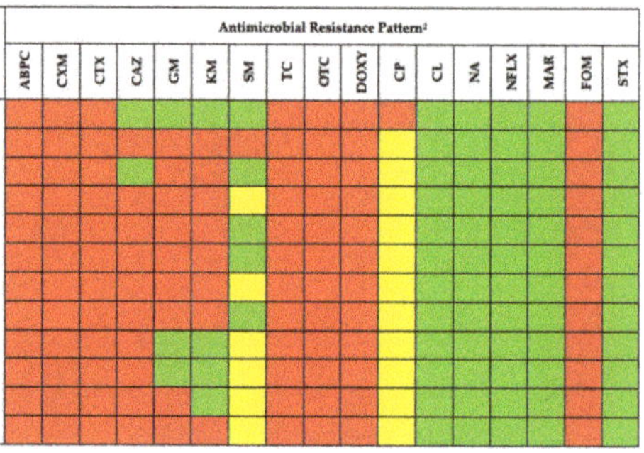

Figure 1. Characteristics of extended-spectrum β-lactamase/AmpC β-lactamase-producing *Klebsiella pneumoniae* (ESBL/AmpC-KP) isolated from Thoroughbred racehorses in Japan. [1] Miho, Miho Training Center, Japan Racing Association, Ibaraki; Ritto, Ritto Training Center, Japan Racing Association, Shiga. [2] ABPC, ampicillin; CXM, cefuroxime; CTX, cefotaxime; CAZ, ceftazidime; GM, gentamicin; KM, kanamycin; SM, streptomycin; TC, tetracycline; OTC, oxytetracycline; DOXY, doxycycline; CP, chloramphenicol; CL, colistin; NA, nalidixic acid; NFLX, norfloxacin; MAR, marbofloxacin; FOM, fosfomycin; STX, trimethoprim/sulfamethoxazole; red, resistance; yellow, intermediate; green, susceptible.

Code of Isolate	Sampling Location [1]	Stable Origin	ESBL/AmpC Phenotype	Sequence Type	β-Lactamase Gene	Other Resistance Gene
JMT68b-KP	Miho	A	ESBL	ST-133	bla_{SHV-1}, $bla_{CTX-M-2}$	*tetA, tetB*
JRT4a-KP	Ritto	B	AmpC	ST-145	bla_{SHV-1}, bla_{CMY}	*strA, strB, tetA, floR*
JRT6a-KP	Ritto	C	AmpC	ST-145	bla_{SHV-1}, bla_{CMY}	*strA, strB, tetA, floR*
JRT6b-KP	Ritto	C	AmpC	ST-145	bla_{SHV-1}, bla_{CMY}	*strA, strB, tetA, floR*
JRT39a-KP	Ritto	D	AmpC	ST-4831	bla_{SHV-1}, bla_{CMY}	*strA, strB, tetA*
JRT39b-KP	Ritto	D	AmpC	ST-145	bla_{SHV-1}, bla_{CMY}	*strA, strB, tetA, floR*
JRT39c-KP	Ritto	D	AmpC	ST-145	bla_{SHV-1}, bla_{CMY}	*strA, strB, tetA, floR*
JRT40a-KP	Ritto	E	AmpC	ST-4830	bla_{SHV-1}, bla_{CMY}	*strA, strB, floR*
JRT40b-KP	Ritto	E	AmpC	ST-4830	bla_{SHV-1}, bla_{CMY}	*strA, strB, tetA, floR*
JRT41a-KP	Ritto	F	AmpC	ST-4830	bla_{SHV-1}, bla_{CMY}	*strA, strB, tetA, floR*
JRT41b-KP	Ritto	F	AmpC	ST-4832	bla_{SHV-1}, bla_{CMY}	*strA, strB, tetA, floR*
JRT79a-KP	Ritto	G	AmpC	ST-145	bla_{SHV-1}, bla_{CMY}	*strA, strB, tetA, floR*

Antimicrobial Resistance Pattern [2] (color-coded cells: R = red/resistance, I = yellow/intermediate, S = green/susceptible):

Code of Isolate	ABPC	CXM	CTX	CAZ	GM	KM	SM	TC	OTC	DOXY	CP	CL	NA	NFLX	MAR	FOM	STX
JMT68b-KP	R	R	R	S	S	S	S	R	R	R	R	S	S	S	S	R	S
JRT4a-KP	R	R	R	R	R	R	R	R	R	R	I	S	S	S	S	R	S
JRT6a-KP	R	R	R	S	R	R	S	R	R	R	I	S	S	S	S	R	S
JRT6b-KP	R	R	R	R	R	R	I	R	R	R	I	S	S	S	S	R	S
JRT39a-KP	R	R	R	R	R	R	S	R	R	R	I	S	S	S	S	R	S
JRT39b-KP	R	R	R	R	R	R	S	R	R	R	I	S	S	S	S	R	S
JRT39c-KP	R	R	R	R	R	R	I	R	R	R	I	S	S	S	S	R	S
JRT40a-KP	R	R	R	R	R	R	S	R	R	R	I	S	S	S	S	R	S
JRT40b-KP	R	R	R	R	S	S	I	R	R	R	I	S	S	S	S	R	S
JRT41a-KP	R	R	R	R	S	S	I	R	R	R	I	S	S	S	S	R	S
JRT41b-KP	R	R	R	R	R	S	I	R	R	R	I	S	S	S	S	R	S
JRT79a-KP	R	R	R	R	R	R	I	R	R	R	I	S	S	S	S	R	S

3.2. Molecular Characteristic of ESBL/AmpC-KP

The presence of SHV-1 β-lactamase genes (non-ESBL) were detected in all (100%; 12/12) ESBL/AmpC-KP isolates, and CTX-M-2 was only detected from one isolate that showed an ESBL phenotype. None of the isolates were positive for chromosomal AmpC genes, whereas CMY, which belonged to the CIT family of pAmpC, was detected in all AmpC phenotype isolates. The *strA, strB,* and *tetA,* genes were detected in nearly all (91.7%; 11/12) ESBL/AmpC-KP isolates, followed by *floR* (75.0%; 9/12) and *tetB,* which was only detected in one isolate (8.3%; 1/12). All isolates were subjected to MLST analysis. As a result, ESBL-KP isolate was identified as ST-133, and four different STs among AmpC-KP isolates, ST-145 (54.5%; 6/11), ST-4830 (27.3% 3/11), ST-4831 (9.1%; 1/11), and ST-4832 (9.1%; 1/11), shared six of the seven loci constituting a single-locus variant (SLV). In these results, AmpC-KP ST-4830, ST-4831, and ST-4832 were termed as new STs. Characteristic ESBL/AmpC-KP and other resistance genes are summarized in Figure 1.

3.3. Conjugation Assay

Conjugation assay was only successful in ESBL-KP ST-133. Horizontal transmission was confirmed by detection of $bla_{CTX-M-2}$ in the transconjugant strain with the frequency of transfer 2×10^{-4} per donor cell.

4. Discussion

In this study, 3.3% of samples from racehorse feces were confirmed as having ESBL/AmpC-KP. Interestingly, 91.7% of total isolates were AmpC producers, which were only isolated from the Ritto Training Center. One isolate (8.3%) was identified as ESBL-KP, derived from the Miho Training Center. ESBL-KP isolated from horses was reported at 0.2% (3/1347) in the Netherlands [13]. In Germany and other European countries, 3.1% (5/160) of ESBL-KP was reported among clinical horse samples [30]. Another study showed that ESBL-KP was isolated from 1.8% (1/55) of foals on admission to hospital, and the shedding rate increased during hospitalization in Israel [31]. The selection of ESBL producers among *Enterobacteriaceae* is expected as the impact of cephalosporin antibiotics used for medical treatment in horses [24].

In our results, all ESBL/AmpC-KP isolates were detected as carrying bla_{SHV-1}, which is resistant to penicillin and early generation cephalosporin but not resistant to third-generation cephalosporin. SHV-1 is mainly reported in *K. pneumoniae* and may be due to the gene encoded SHV-1, which was located on the chromosome of this species. SHV-1 β-lactamase has also been reported for up to 20% of plasmid-mediated ampicillin *K. pneumoniae* [32]. Our study also confirmed that ESBL-KP isolate was carried the $bla_{CTX-M-2}$ gene. CTX-M-2-producing *E. coli* were also detected from the same horse feces sample (data not shown), as reported in our previous study [33]. Conjugation assay showed that $bla_{CTX-M-2}$ was transferred with the frequency of transfer 2×10^{-4} per donor cell. This finding suggests that horizontal transmission among bacterial species in horse intestine occurred. In Japan, CTX-M-2-producing *K. pneumoniae* have been confirmed in dogs [7], humans [34,35], and broiler chickens [36]. In addition, conjugative plasmids carrying $bla_{CTX-M-2}$ have been reported in *K. pneumoniae* isolated from dairy cows with clinical mastitis [37]. In contrast to ESBL-KP, the presence of AmpC-KP in horses is less well documented, but our study identified them as a dominant β-lactamase producer.

The screening test for detection of AmpC-producing bacteria can be performed by the same protocol for ESBL screening test, and multiplex PCR has been developed to identify pAmpC [17]. All AmpC phenotype isolates in our study contained bla_{CMY} belonging to the bla_{CIT} type of the pAmpC gene. CMY-2 is prevalent among AmpC enzymes in the animal sector [38]. None of the pAmpC-KP isolates were conjugative under our experimental conditions. To our knowledge, no previous studies have been published describing the rate of *K. pneumoniae* carrying pAmpC isolated from horses. Plasmid-mediated AmpC has been reported worldwide from enterobacteria not predicted to produce AmpC β-lactamases [12]. In equine medicine, previous studies have shown that pAmpC genes belonging to bla_{CMY-2} were detected from extended-spectrum cephalosporin-resistant (ESCR) *E. coli* isolated from diseased horses in the Netherlands (0.1%; 1/1347) and the United Kingdom (3.8%; 2/52) [13,39]. The bla_{CMY} was also identified from *Salmonella* spp. isolated from horses in the United States and Ireland [8]. The bla_{EBC} (5.8%; 3/52) identified from ESCR *E. coli* has been reported in the United Kingdom [39]. Plasmid-mediated AmpC-KP has been isolated from dogs and/or cats in South Korea [10], China [40,41], Japan [7], Switzerland [42], and Italy [3], and most of these belong to the CMY and DHA groups. In this work, no AmpC-KP isolates were also confirmed as ESBL producers, and vice versa. This might be related to the antimicrobials used in the treatment of animals [43]. In a previous study, CTX-M-2- and CMY-2-producing *E. coli* were reported in broiler chickens in Japan [44]. In addition, the susceptibility to carbapenems could be decreased by combination of AmpC production and porin deficiency [18]. Nevertheless, no ESBL/AmpC-KP showed activity as carbapenemase producers in this study.

ESBL and pAmpC-producing bacteria mostly have co-resistance with other antimicrobials [3,38]. The ESBL/AmpC genes are frequently located on an MDR plasmid, which plays a key role in their dissemination [45]. Our results showed the occurrence of MDR ESBL/AmpC-KP isolated from horses (3.3%; 7/212) was lower than from dogs and cats (30.1%; 31/103) in Japan [7]. Most MDR ESBL/AmpC-KP isolates showed co-resistance with aminoglycoside (*strA*- and *strB*-positive), tetracycline (*tetA*- and/or *tetB*-positive), and FOM. Only ESBL-KP isolates showed resistance to CP, but the *floR* gene, which is responsible for CP resistance, was detected in most CP non-susceptible AmpC-KP isolates. Similar to our results, MDR ESBL/AmpC-KP against aminoglycosides, tetracyclines, and amphenicol-mediated *strA/B*, *tet*, and *cat* genes have also been confirmed from dogs and cats in Italy [3]. Co-selection, when using antimicrobials other than ESCs for therapy, may maintain the existence of MDR ESBL/AmpC-producing bacteria in animals [38]. Treatment options for MDR ESBL/AmpC-KP infection might be limited when considering that several clinical cases have been reported from this species in horses.

MLST analysis showed that *K. pneumoniae* ST-133 was identified as an ESBL producer in this study. Previously, ESBL-KP ST-133 has been reported in humans in Japan [46]. Four different STs of AmpC-KP (ST-145, ST-4830, ST-4831, and ST-4832) in this study have not been reported between humans and animals in Japan. AmpC-KP ST-145 and three new STs, which are SLV of ST-145, were only distributed at the JRA Ritto Training Center. Further investigation is needed to confirm whether the dissemination of ESBL/AmpC-KP occurred inside or outside the training center.

5. Conclusions

In conclusion, this is the first study that has shown *K. pneumoniae* carrying MDR pAmpC isolated from racehorses. Interestingly, our results showed that the percentage of pAmpC-KP is higher than ESBL-KP, as compared with other previous reports. Dissemination of MDR ESBL/AmpC-KP through fecal material in the training centers requires special attention among the racehorse community, as indirect transmission may occur in the environment. Risk of infection by MDR ESBL/AmpC-KP may occur in people who work in close contact with racehorses (e.g., veterinarians, caretakers, and owners).

Author Contributions: Conceptualization, E.S. and R.U.; methodology, E.S., W.S., and M.T.H.; software, E.S.; validation, E.S., R.U., and M.S.; formal analysis, E.S. and R.U.; investigation, E.S.; resources, E.S. and R.U.; data curation, E.S.; writing—original draft preparation, E.S.; writing—review and editing, E.S., R.U., and M.S.; visualization, E.S.; supervision, R.U. and M.S.; project administration, R.U.; funding acquisition, R.U. All authors have read and agreed to the published version of the manuscript.

Funding: This research received no external funding.

Acknowledgments: The authors gratefully thank the Japan Ricing Association for providing the fecal samples for use in the present study. We are grateful to the team of curators of the Institut Pasteur MLST and whole genome MLST databases for curating the data and making them publicly available at http://bigsdb.pasteur.fr.

Conflicts of Interest: The authors declare no conflict of interest.

References

1. Trigo da Roza, F.; Couto, N.; Carneiro, C.; Cunha, E.; Rosa, T.; Magalhães, M.; Tavares, L.; Novais, Â.; Peixe, L.; Rossen, J.W.; et al. Commonality of multidrug-resistant *Klebsiella pneumoniae* ST348 isolates in horses and humans in Portugal. *Front. Microbiol.* **2019**, *10*, 1657. [CrossRef] [PubMed]

2. Estell, K.E.; Young, A.; Kozikowski, T.; Swain, E.A.; Byrne, B.A.; Reilly, C.M.; Kass, P.H.; Aleman, M. Pneumonia caused by *Klebsiella* spp. in 46 horses. *J. Vet. Intern. Med.* **2016**, *30*, 314–321. [CrossRef] [PubMed]

3. Donati, V.; Feltrin, F.; Hendriksen, R.S.; Svendsen, C.A.; Cordaro, G.; García-Fernández, A.; Lorenzetti, S.; Lorenzetti, R.; Battisti, A.; Franco, A. Extended-spectrum-beta-lactamases, AmpC beta-lactamases and plasmid mediated quinolone resistance in *Klebsiella* spp. from companion animals in Italy. *PLoS ONE* **2014**, *9*, e90564. [CrossRef] [PubMed]

4. Satué, K.; Gardon, J.C. Infection and infertility in mares. In *Genital Infections and Infertility*; Darwish, A., Ed.; IntechOpen: London, UK, 2016; Available online: https://www.intechopen.com/books/genital-infections-and-infertility/infection-and-infertility-in-mares (accessed on 10 February 2020).

5. Lam, M.M.C.; Wyres, K.L.; Duchêne, S.; Wick, R.R.; Judd, L.M.; Gan, Y.H.; Hoh, C.H.; Archuleta, S.; Molton, J.S.; Kalimuddin, S.; et al. Population genomics of hypervirulent *Klebsiella pneumoniae* clonal-group 23 reveals early emergence and rapid global dissemination. *Nat. Commun.* **2018**, *9*, 2703. [CrossRef]

6. Magdesian, K.G. Update on common antimicrobials. In *Current Therapy in Equine Medicine*, 6th ed.; Robinson, N.E., Sprayberry, K.A., Eds.; Saunders Elsevier: St. Louis, MI, USA, 2009; pp. 10–14.

7. Harada, K.; Shimizu, T.; Mukai, Y.; Kuwajima, K.; Sato, T.; Usui, M.; Tamura, Y.; Kimura, Y.; Miyamoto, T.; Tsuyuki, Y.; et al. Phenotypic and molecular characterization of antimicrobial resistance in *Klebsiella* spp.

isolates from companion animals in Japan: Clonal dissemination of multidrug-resistant extended-spectrum β-lactamase-producing *Klebsiella pneumoniae*. *Front. Microbiol.* **2016**, *7*, 1021. [CrossRef]

8. Ewers, C.; Bethe, A.; Semmler, T.; Guenther, S.; Wieler, L.H. Extended-spectrum β-lactamase-producing and AmpC-producing *Escherichia coli* from livestock and companion animals, and their putative impact on public health: A global perspective. *Clin. Microbiol. Infect.* **2012**, *18*, 646–655. [CrossRef]

9. Rupp, M.E.; Fey, P.D. Extended spectrum β-lactamase (ESBL)-producing *Enterobacteriaceae*: Considerations for diagnosis, prevention and drug treatment. *Drugs* **2003**, *63*, 353–365. [CrossRef]

10. Hong, J.S.; Song, W.; Park, H.M.; Oh, J.Y.; Chae, J.C.; Shin, S.; Jeong, S.H. Clonal spread of extended-spectrum cephalosporin-resistant *Enterobacteriaceae* between companion animals and humans in South Korea. *Front. Microbiol.* **2019**, *10*, 1371. [CrossRef]

11. Tepeli, S.Ö.; Demirel Zorba, N.N. Frequency of extended-spectrum β-lactamase (ESBL)- and AmpC β-lactamase-producing *Enterobacteriaceae* in a cheese production process. *J. Dairy. Sci.* **2018**, *101*, 2906–2914. [CrossRef]

12. Jacoby, G.A. AmpC β-lactamases. *Clin. Microbiol. Rev.* **2009**, *22*, 161–182. [CrossRef]

13. Vo, A.T.; van Duijkeren, E.; Fluit, A.C.; Gaastra, W. Characteristics of extended-spectrum cephalosporin-resistant *Escherichia coli* and *Klebsiella pneumoniae* isolates from horses. *Vet. Microbiol.* **2007**, *124*, 248–255. [CrossRef] [PubMed]

14. Weese, J.S. Antimicrobial use and antimicrobial resistance in horses. *Equine Vet. J.* **2015**, *47*, 747–749. [CrossRef] [PubMed]

15. Doi, Y.; Iovleva, A.; Bonomo, R.A. The ecology of extended-spectrum β-lactamases (ESBLs) in the developed world. *J. Travel. Med.* **2017**, *24*, S44–S51. [CrossRef] [PubMed]

16. Melo, L.C.; Oresco, C.; Leigue, L.; Netto, H.M.; Melville, P.A.; Benites, N.R.; Saras, E.; Haenni, M.; Lincopan, N.; Madec, J.Y. Prevalence and molecular features of ESBL/pAmpC-producing *Enterobacteriaceae* in healthy and diseased companion animals in Brazil. *Vet. Microbiol.* **2018**, *221*, 59–66. [CrossRef]

17. Thomson, K.S. Extended-spectrum-β-lactamase, AmpC, and Carbapenemase issues. *J. Clin. Microbiol.* **2010**, *48*, 1019–1025. [CrossRef]

18. Schmiedel, J.; Falgenhauer, L.; Domann, E.; Bauerfeind, R.; Prenger-Berninghoff, E.; Imirzalioglu, C.; Chakraborty, T. Multiresistant extended-spectrum β-lactamase-producing *Enterobacteriaceae* from humans, companion animals and horses in central Hesse, Germany. *BMC. Microbiol.* **2014**, *14*, 187. [CrossRef]

19. Dandachi, I.; Chabou, S.; Daoud, Z.; Rolain, J.M. Prevalence and emergence of extended-spectrum cephalosporin-, carbapenem- and colistin-resistant gram negative bacteria of animal origin in the Mediterranean Basin. *Front. Microbiol.* **2018**, *9*, 2299. [CrossRef]

20. EUCAST Guidelines for Detection of Resistance Mechanisms and Specific Resistances of Clinical and/or Epidemiological Importance. Available online: http://www.eucast.org/fileadmin/src/media/PDFs/EUCAST_files/Resistance_mechanisms/EUCAST_detection_of_resistance_mechanisms_v1.0_20131211.pdf (accessed on 24 November 2016).

21. CLSI. *Performance Standards for Antimicrobial Susceptibility Testing*, 26th ed.; CLSI Suppl. M100S; Clinical and Laboratory Standards Institute: Wayne, PA, USA, 2016.

22. Magiorakos, A.P.; Srinivasan, A.; Carey, R.B.; Carmeli, Y.; Falagas, M.E.; Giske, C.G.; Harbarth, S.; Hindler, J.F.; Kahlmeter, G.; Olsson-Liljequist, B.; et al. Multidrug-resistant, extensively drug-resistant and pandrug-resistant bacteria: An international expert proposal for interim standard definitions for acquired resistance. *Clin. Microbiol. Infect.* **2012**, *18*, 268–281. [CrossRef]

23. Yamazaki, W.; Kumeda, Y.; Uemura, R.; Misawa, N. Evaluation of a loop-mediated isothermal amplification assay for rapid and simple detection of *Vibrio parahaemolyticus* in naturally contaminated seafood samples. *Food Microbiol.* **2011**, *28*, 1238–1241. [CrossRef]

24. Sukmawinata, E.; Sato, W.; Mitoma, S.; Kanda, T.; Kusano, K.; Kambayashi, Y.; Sato, T.; Ishikawa, Y.; Goto, Y.; Uemura, R.; et al. Extended-spectrum β-lactamase-producing *Escherichia coli* isolated from healthy Thoroughbred racehorses in Japan. *J. Equine. Sci.* **2019**, *30*, 47–53. [CrossRef]

25. Pérez-Pérez, F.J.; Hanson, N.D. Detection of plasmid-mediated AmpC β-lactamase genes in clinical isolates by using multiplex PCR. *J. Clin. Microbiol.* **2002**, *40*, 2153–2162. [CrossRef] [PubMed]

26. Dierikx, C.M.; van Duijkeren, E.; Schoormans, A.H.; van Essen-Zandbergen, A.; Veldman, K.; Kant, A.; Huijsdens, X.W.; van der Zwaluw, K.; Wagenaar, J.A.; Mevius, D.J. Occurrence and characteristics of

extended-spectrum-β-lactamase- and AmpC-producing clinical isolates derived from companion animals and horses. *J. Antimicrob. Chemother.* **2012**, *67*, 1368–1374. [CrossRef] [PubMed]

27. Yamamoto, S.; Iwabuchi, E.; Hasegawa, M.; Esaki, H.; Muramatsu, M.; Hirayama, N.; Hirai, K. Prevalence and molecular epidemiological characterization of antimicrobial-resistant *Escherichia coli* isolates from Japanese black beef cattle. *J. Food. Prot.* **2013**, *76*, 394–404. [CrossRef] [PubMed]

28. Karczmarczyk, M.; Abbott, Y.; Walsh, C.; Leonard, N.; Fanning, S. Characterization of multidrug-resistant *Escherichia coli* isolates from animals presenting at a university veterinary hospital. *Appl. Environ. Microbiol.* **2011**, *77*, 7104–7112. [CrossRef] [PubMed]

29. Diancourt, L.; Passet, V.; Verhoef, J.; Grimont, P.A.; Brisse, S. Multilocus sequence typing of *Klebsiella pneumoniae* nosocomial isolates. *J. Clin. Microbiol.* **2005**, *43*, 4178–4182. [CrossRef]

30. Ewers, C.; Stamm, I.; Pfeifer, Y.; Wieler, L.H.; Kopp, P.A.; Schønning, K.; Prenger-Berninghoff, E.; Scheufen, S.; Stolle, I.; Günther, S.; et al. Clonal spread of highly successful ST15-CTX-M-15 *Klebsiella pneumoniae* in companion animals and horses. *J. Antimicrob. Chemother.* **2014**, *69*, 2676–2680. [CrossRef]

31. Shnaiderman-Torban, A.; Paitan, Y.; Arielly, H.; Kondratyeva, K.; Tirosh-Levy, S.; Abells-Sutton, G.; Navon-Venezia, S.; Steinman, A. Extended-spectrum β-lactamase-producing Enterobacteriaceae in hospitalized neonatal foals: Prevalence, risk factors for shedding and association with infection. *Animals* **2019**, *9*, 600. [CrossRef]

32. Shaikh, S.; Fatima, J.; Shakil, S.; Rizvi, S.M.; Kamal, M.A. Antibiotic resistance and extended spectrum beta-lactamases: Types, epidemiology and treatment. *Saudi J. Biol. Sci.* **2015**, *22*, 90–101. [CrossRef]

33. Sukmawinata, E.; Uemura, R.; Sato, W.; Mitoma, S.; Kanda, T.; Sueyoshi, M. IncI1 plasmid associated with bla_{CTX-M-2} transmission in ESBL-producing *Escherichia coli* isolated from healthy Thoroughbred racehorse, Japan. *Antibiotics* **2020**, *9*, 70. [CrossRef]

34. Hawkey, P.M. Prevalence and clonality of extended-spectrum β-lactamases in Asia. *Clin. Microbiol. Infect.* **2008**, *14*, 159–165. [CrossRef]

35. Chong, Y.; Shimoda, S.; Yakushiji, H.; Ito, Y.; Miyamoto, T.; Kamimura, T.; Shimono, N.; Akashi, K. Community spread of extended-spectrum β-lactamase-producing *Escherichia coli*, *Klebsiella pneumoniae* and *Proteus mirabilis*: A long-term study in Japan. *J. Med. Microbiol.* **2013**, *62*, 1038–1043. [CrossRef] [PubMed]

36. Hiroi, M.; Yamazaki, F.; Harada, T.; Takahashi, N.; Iida, N.; Noda, Y.; Yagi, M.; Nishio, T.; Kanda, T.; Kawamori, F.; et al. Prevalence of extended-spectrum β-lactamase-producing *Escherichia coli* and *Klebsiella pneumoniae* in food-producing animals. *J. Vet. Med. Sci.* **2012**, *74*, 189–195. [CrossRef] [PubMed]

37. Saishu, N.; Ozaki, H.; Murase, T. CTX-M-type extended-spectrum β-lactamase-producing *Klebsiella pneumoniae* isolated from cases of bovine mastitis in Japan. *J. Vet. Med. Sci.* **2014**, *76*, 1153–1156. [CrossRef] [PubMed]

38. Madec, J.Y.; Haenni, M.; Nordmann, P.; Poirel, L. Extended-spectrum β-lactamase/AmpC- and carbapenemase-producing Enterobacteriaceae in animals: A threat for humans? *Clin. Microbiol. Infect.* **2017**, *23*, 826–833. [CrossRef] [PubMed]

39. Bortolami, A.; Zendri, F.; Maciuca, E.I.; Wattret, A.; Ellis, C.; Schmidt, V.; Pinchbeck, G.; Timofte, D. Diversity, virulence, and clinical significance of extended-spectrum β-lactamase- and pAmpC-producing *Escherichia coli* from companion animals. *Front. Microbiol.* **2019**, *10*, 1260. [CrossRef] [PubMed]

40. Ma, J.; Zeng, Z.; Chen, Z.; Xu, X.; Wang, X.; Deng, Y.; Lü, D.; Huang, L.; Zhang, Y.; Liu, J.; et al. High prevalence of plasmid-mediated quinolone resistance determinants qnr, aac(6')-Ib-cr, and qepA among ceftiofur-resistant Enterobacteriaceae isolates from companion and food-producing animals. *Antimicrob. Agents. Chemother.* **2009**, *53*, 519–524. [CrossRef]

41. Liu, Y.; Yang, Y.; Chen, Y.; Xia, Z. Antimicrobial resistance profiles and genotypes of extended-spectrum β-lactamase- and AmpC β-lactamase-producing *Klebsiella pneumoniae* isolated from dogs in Beijing, China. *J. Glob. Antimicrob. Resist.* **2017**, *10*, 219–222. [CrossRef]

42. Wohlwend, N.; Endimiani, A.; Francey, T.; Perreten, V. Third-generation-cephalosporin-resistant *Klebsiella pneumoniae* isolates from humans and companion animals in Switzerland: Spread of a DHA-producing sequence type 11 clone in a veterinary setting. *Antimicrob. Agents. Chemother.* **2015**, *59*, 2949–2955. [CrossRef]

43. Shiraki, Y.; Shibata, N.; Doi, Y.; Arakawa, Y. *Escherichia coli* producing CTX-M-2 β-lactamase in cattle, Japan. *Emerg. Infect. Dis.* **2004**, *10*, 69–75. [CrossRef]

44. Kameyama, M.; Chuma, T.; Yabata, J.; Tominaga, K.; Iwata, H.; Okamoto, K. Prevalence and epidemiological relationship of CMY-2 AmpC β-lactamase and CTX-M extended-spectrum β-lactamase-producing *Escherichia coli* isolates from broiler farms in Japan. *J. Vet. Med. Sci.* **2013**, *75*, 1009–1015. [CrossRef]

45. De Lagarde, M.; Larrieu, C.; Praud, K.; Schouler, C.; Doublet, B.; Sallé, G.; Fairbrother, J.M.; Arsenault, J. Prevalence, risk factors, and characterization of multidrug resistant and extended spectrum β-lactamase/AmpC β-lactamase producing *Escherichia coli* in healthy horses in France in 2015. *J. Vet. Intern. Med.* **2019**, *33*, 902–911. [CrossRef] [PubMed]

46. *Klebsiella* PasteurMLST Database. Available online: https://bigsdb.pasteur.fr/cgi-bin/bigsdb/bigsdb.pl?db=pubmlst_klebsiella_isolates (accessed on 8 January 2020).

Article

Broad-Spectrum Cephalosporin-Resistant *Klebsiella* spp. Isolated from Diseased Horses in Austria

Igor Loncaric [1,*], Adriana Cabal Rosel [2], Michael P. Szostak [1], Theresia Licka [3], Franz Allerberger [2], Werner Ruppitsch [2] and Joachim Spergser [1]

[1]   Institute of Microbiology, University of Veterinary Medicine, 1210 Vienna, Austria; michael.szostak@vetmeduni.ac.at (M.P.S); Joachim.spergser@vetmeduni.ac.at (J.S.)

[2]   Institute of Medical Microbiology and Hygiene, Austrian Agency for Health and Food Safety, 1090 Vienna, Austria; adriana.cabal-rosel@ages.at (A.C.R.); franz.allerberger@ages.at (F.A.); werner.ruppitsch@ages.at (W.R.)

[3]   Clinical Unit of Equine Surgery, University of Veterinary Medicine, 1210 Vienna, Austria; theresia.licka@vetmeduni.ac.at

*   Correspondence: igor.loncaric@vetmeduni.ac.at; Tel.: +43-1-250772115

Received: 30 January 2020; Accepted: 15 February 2020; Published: 20 February 2020

Simple Summary: Broad-spectrum cephalosporin-resistant *Klebsiella pneumoniae* is considered as a serious problem for public human health. To date, only a few broad-spectrum cephalosporin-resistant *Klebsiella* have been isolated from horses. Considering the zoonotic potential of the *Klebsiella* spp., and the close relationship between man and horse, this study intended to generate data on the genetic background of broad-spectrum cephalosporin-resistant *Klebsiella* spp. isolated from horses in Austria. Overall, samples isolated between 2012 and 2019 from 1541 horses underwent bacteriological testing, resulting in 51 specimens tested positive for *Klebsiella* ssp. Antimicrobial susceptibility tests revealed that seven *Klebsiella* ssp. isolates were not only cefotaxime-resistant but also showed resistance against other classes of antibiotics so that they were considered to be multidrug-resistant. Data from whole genome sequencing and mating experiments strongly suggest that the majority of antibiotic resistance genes is encoded on plasmids in these seven multidrug-resistant *Klebsiella* ssp. Considering the potential threat when commensal *Klebsiella* inhabiting a healthy human gut acquire new antibiotic resistances due to the exchange of plasmids with multidrug-resistant *Klebsiella* ssp. from horses, further monitoring of horses and other domestic animals for the presence of broad-spectrum cephalosporin-resistant *Klebsiella*, not only in Austria but worldwide is therefore advisable.

Abstract: The aim of the present study was to investigate the diversity of broad-spectrum cephalosporin-resistant *Klebsiella* spp. isolated from horses in Austria that originated from diseased horses. A total of seven non-repetitive cefotaxime-resistant *Klebsiella* sp. isolates were obtained during diagnostic activities from autumn 2012 to October 2019. Antimicrobial susceptibility testing was performed. The isolates were genotyped by whole-genome sequencing (WGS). Four out of seven *Klebsiella* isolates were identified as *K. pneumoniae*, two as *K. michiganensis* and one as *K. oxytoca*. All isolates displayed a multi-drug resistant phenotype. The detection of resistance genes reflected well the phenotypic resistance profiles of the respective isolates. All but one isolate displayed the extended-spectrum β-lactamases (ESBL) phenotype and carried CTX-M cefotaximases, whereas one isolate displayed an ESBL and AmpC phenotype and carried cephamycinase (CMY)-2 and sulfhydryl variable (SHV)-type b and Temoniera (TEM) β-lactamases. Among *Klebsiella pneumoniae* isolates, for different sequence types (ST) could be detected (ST147, ST307, ST1228, and a new ST4848). Besides resistance genes, a variety of virulence genes, including genes coding for yersiniabactin were detected. Considering the high proximity between horses and humans, our results undoubtedly identified a public health issue. This deserves to be also monitored in the years to come.

Keywords: AmpC; ESBL; *Klebsiella pneumoniae*; antibiotic-resistance; β-lactamases; horses

1. Introduction

Among the member of the genus *Klebsiella*, broad-spectrum cephalosporin-resistant *Klebsiella* (*K.*) *pneumoniae* is frequently associated with severe nosocomial infections in humans, and due to its antibiotic-resistant traits, infections leave limited therapeutic options [1,2]. In early 2017, the World Health Organization (WHO) listed carbapenem-resistant and 3rd generation cephalosporin-resistant *Enterobacteriales* (including, e.g., *K. pneumoniae*, *Escherichia coli*, *Enterobacter* spp., *Serratia* spp., *Proteus* spp., *Providencia* spp., *Morganella* spp.) "Priority 1: Critical group" bacterial pathogens. These bacteria are in focus on the discovery and development of new antibiotics [2,3].

Today, broad-spectrum cephalosporin-resistant *K. pneumoniae* is recognized as a serious public health problem in human medicine [4,5]. Contrarily, there is still a scarcity of information on broad-spectrum cephalosporin-resistant *K. pneumoniae* and members of the genus *Klebsiella* isolated from horses and other domestic animals. To date, only a few equine broad-spectrum cephalosporin-resistant *Klebsiella* have been isolated and characterized [4,6–10]. Recent studies reported that some of the characterized resistant *K. pneumoniae* isolates of equine origin were human-associated multidrug-resistant (MDR) *K. pneumoniae* [4,8]. At present, there are no published data on the genetic background of broad-spectrum cephalosporin-resistant *Klebsiella* spp. isolated from horses in Austria. Therefore, there is a need to generate such data to understand the molecular epidemiology of these particular pathogens.

In the present study, we have characterized a collection of equine broad-spectrum cephalosporin-resistant *Klebsiella* sp. from clinical samples by multiphasic approach, including whole-genome sequencing (WGS).

2. Materials and Methods

At the Institute of Microbiology, University of Veterinary Medicine, Vienna, approximately 350 susceptibility tests are performed on clinical isolates from horses each year. During the study period (2012 until October 2019), samples of 1541 horses underwent bacteriological testing. In 51 specimens, *Klebsiella* sp. was detected, wherefrom a total of seven non-repetitive cefotaxime-resistant isolates, which were identified to the species level by matrix-assisted laser desorption/ionization-time-of-flight (MALDI-TOF) mass spectrometry (Bruker Daltonik, Heidelberg, Germany), and were further analyzed. They originated from lavage (isolates 1505 and 2826), wound (isolates 2668 and 2742), fistula (isolate 1635), trachea (isolate 2341b), and feces (isolate 4545). All isolates were stored in glycerol stocks at −80 °C. All samples originated from non-food producing horses. All these clinical samples were received from third parties and, therefore, not subject to reporting obligations of the Ethics and Animal Welfare Commission of the University of Veterinary Medicine in Vienna.

Antimicrobial susceptibility testing was performed by agar disk-diffusion according to standards of the Clinical and Laboratory Standards Institute (CLSI) [11]. *Escherichia coli* ATCC® 25922 served as quality control strains. The following antimicrobials were used: cefotaxime, ceftazidime, aztreonam, imipenem, meropenem, gentamicin, amikacin, tobramycin, ciprofloxacin, trimethoprim-sulfamethoxazole, tetracycline, chloramphenicol, and fosfomycin (Becton Dickinson, Heidelberg, Germany). In addition, isolates were checked for extended-spectrum β-lactamase (ESBL) production by ESBL-test via agar disk diffusion [11]. Furthermore, cefoxitin (30 µg) was added to this test to detect AmpC phenotypes.

Whole-genome sequencing (WGS) was performed by isolating and sequencing bacterial DNA, as previously described [12]. *De novo* assembly of raw reads, whole genome sequencingt (WGS) data analysis, including multi-locus sequence typing (MLST) and core genome multi-locus sequence-based typing (cgMLST), were performed, as previously described [13,14].

Species identification was conducted with the JSpecies workspace using the ANIb (average nucleotide identity via Basic Local Alignment Search Tool (BLAST) analysis tool [15]. The identification of acquired resistance genes and chromosomal mutations was performed using the Comprehensive Antibiotic Resistance Database (CARD; https://card.mcmaster.ca/home) [16], as well as ResFinder 3.2 (https://cge.cbs.dtu.dk/services/ResFinder/) [17] were used. eBURST (Based Upon Related Sequence Types) analysis (a plugin at https://bigsdb.pasteur.fr/) was conducted to identify clonal complexes (CCs), defined as groups of two or more independent isolates sharing identical alleles at six or more loci.

The presence of plasmids was determined using PlasmidFinder 1.3 available from the Center for Genomic Epidemiology web server (http://www.genomicepidemiology.org/) [18]. Probability Prediction of the location of a given antibiotic resistance gene was achieved by applying mlplasmids trained on *K. pneumonia* [19]. Posterior probability scores >0.7 and a minimum contig length of 1000 bp indicate that a given contig is plasmid-derived.

Mating experiments were conducted by conjugation as well as transformation, as previously described [20]. Variable regions of class 1 and class 2 integrons were determined by PCR [20]. The quinolone resistance-determining regions (QRDR) of *gyrA* and *parC* in ciprofloxacin-resistant isolates were amplified by PCR and sequenced [21].

The presence of virulence genes was examined by using the virulence allele library from the Institute Pasteur BIGSdb database for *K. pneumoniae* (http://bigsdb.pasteur.fr/klebsiella).

This whole-genome shotgun project has been deposited in DDBJ/EMBL/GenBank under the project number PRJNA600879. Raw sequence data for each strain were deposited under Sequence Read Archive (SRA) accession numbers SRR10899218 to SRR10899224.

3. Results

Four out of seven cefotaxime-resistant *Klebsiella* isolates were identified as *K. pneumoniae*, two as *K. michiganensis*, and one as *K. oxytoca* (Table 1). All but one isolate displayed the ESBL phenotype, whereas one isolate displayed an ESBL and AmpC phenotype. Besides cefotaxime, all *K. pneumoniae* isolates were resistant to ceftazidime, and one isolate additionally to aztreonam. All examined isolates were resistant to gentamicin and tobramycin. None of the analyzed isolates was resistant to carbapenems and amikacin. Five isolates were resistant to tetracycline, doxycycline, and chloramphenicol, whereas six were resistant to trimethoprim-sulfamethoxazole. All *K. pneumoniae* isolates were resistant to ciprofloxacin, and one isolate to fosfomycin (Table 1). Hence, all examined isolates were considered to be multidrug-resistant [22]. The detection of resistance genes reflected well the phenotypic resistance profiles of the examined isolates (Table 1). In two ciprofloxacin-resistant *K. pneumoniae* isolates, beside fluoroquinolone resistance genes *oqxA*, *oqxB*, *qrnB1*, and *aac(6')-Ib-cr*, mutations in the quinolone resistance-determining regions (QRDRs) of the genes *gyrA* and *parC* were observed (Table 1). Three isolates, both *K. michiganensis* isolates and the *K. oxytoca* isolate contained a class 1 integron with a variable part of ca. 1.7 kb in size, which harbored an *aadA5* and a *dfr17* cassette.

In total, ten different plasmids IncFIA(HI1), IncFIB(K), IncFIB(pHCM2), IncHI1A, IncHI1B(R27), IncI1, IncN, IncQ1, IncR, and Col440l were identified (Table 2). They shared between 92.11 and 100% DNA similarity with corresponding reference sequences. A *K. michiganensis* isolate and two *K. pneumoniae* isolates carried IncFIA(HI1), IncFIB(pHCM2), IncHI1A, IncHI1B(R27), and IncQ1. The *K. oxytoca* isolate carried IncI1 and IncN, whereas a *K. pneumoniae* carried IncN and IncR and another *K. michiganensis* IncFIB(K). According to mlplasmids analyses, the majority of resistance genes might be located on plasmids, especially all *bla*$_{CTX}$, *bla*$_{TEM}$, and *bla*$_{OXA}$ genes as well as all detected genes for resistance against aminoglycosides, trimethoprim/sulfamethoxazole, or chloramphenicol (Table 1).

Among virulence factors, *K. pneumoniae* type 3 fimbriae encoded by *mrk* operon genes as well *iutA* (aerobactin siderophores receptor) were detected in all *K. pneumoniae* isolates, whereas genes coding for yersiniabactin (*ybt*) were detected in only one isolate (Table 3).

Table 1. Characteristics of seven examined cefotaxime-resistant *Klebsiella* isolates.

		1505		1635		2341b		2668		2742		2826		4545	
ST [1] CT [3] Origin		*K. michiganensis* n.a. [2] n.a. Lavage	PPP [4]	*K. oxytoca* n.a. n.a. Fistula	PPP	*K. pneumoniae* ST4848 CT4643 Trachea	PPP	*K. pneumoniae* ST1228 CT4644 Wound	PPP	*K. pneumoniae* ST147 ST1202 Wound	PPP	*K. michiganensis* n.a. n.a. Lavage	PPP	*K. pneumoniae* ST307 CT4645 Feces	PPP
β-lactamas	P [5]	CTX		CTX		CTX, CAZ, FOX		CTX, CAZ		CTX, CAZ, ATM		CTX		CTX, CAZ	
	G [6]	**$bla_{CTX-M-1}$** [7] $bla_{OXY-4-1}$	0.745 0.009	$bla_{CTX-M-1}$ $bla_{OXY-2-7}$ bla_{TEM-1B}	0.710 0.003 0.891	**bla_{CMY2}** bla_{SHV} bla_{TEM-1B}	no ppp 0.003 0.965	**$bla_{CTX-M-1}$** **bla_{SHV-11}** bla_{TEM-1B}	0.760 0.001 no ppp	$bla_{CTX-M-15}$ bla_{OXA-1} bla_{SHV-11} bla_{TEM-1B}	0.973 0.961 0.001 0.973	$bla_{CTX-M-1}$ $bla_{OXY-4-1}$	0.753 0.004	$bla_{CTX-M-15}$ bla_{OXA-1} bla_{SHV-28} bla_{TEM-1B}	0.976 0.961 0.002 0.776
Aminoglycosides	P	GEN, TOB		GEN, TOB		GEN, TOB		GEN, TOB		GEN, TOB		GEN, TOB		GEN, TOB	
	G	*aac(3)-IId* ***aadA5*** [8] *aph(3″)-Ib* *aph(3′)-Ia* *aph(6)-Id*	0.854 0.993 0.896 no ppp 0.896	*aac(3)-IId* ***aadA5*** *aph(3″)-Ib* *aph(6)-Id*	0.909 0.955 0.895 0.895	*aac(3)-IIa* *aph(3″)-Ib*	0.956 0.956	*aac(3)-IId*	0.92	*aac(3)-IIa* *aac(6′)-Ib-cr* *aph(3″)-Ib* *aph(6)-Id*	0.963 0.961 0.973 0.973	*aac(3)-IId* ***aadA5*** *aph(3″)-Ib* *aph(3′)-Ia* *aph(6)-Id*	0.921 0.986 0.895 no ppp 0.895	*aac(3)-IIa* *aac(6′)-Ib-cr* *aph(3″)-Ib* *aph(6)-Id*	0.992 0.961 0.976 0.976
Tetracyclines	P	TET, DOX		TET, DOX						TET, DOX		TET, DOX		TET, DOX	
	G	*tet*(B)	0.690	*tet*(B)	0.545					*tet*(A)	0.029	*tet*(B)	0.709	*tet*(A)	0.946
Chloramphenicol	P	CHL		CHL						CHL		CHL		CHL	
	G	*catA1*	0.993	*catA1*	0.945					*catB3*	0.961	*catA1*	0.986	*catB3*	0.961
Trimethoprim/ sulfamethoxazole	P	SXT		SXT		SXT				SXT		SXT		SXT	
	G	*sul1* *sul2* ***dfrA17***	0.993 0.896 0.993	*sul1* *sul2* ***dfrA17***	0.955 0.895 0.955	*sul2* *dfrA14*	0.956 0.876			*sul2* *dfrA14*	0.973 0.968	*sul1* *sul2* ***dfrA17***	0.986 0.895 0.986	*sul2* *dfrA14*	0.976 0.957

Table 1. *Cont.*

		1505	1635	2341b		2668		2742		2826		4545	
Fosfomycin	P					FOS		FOS				FOS	
	G					*fosA*	0.001	*fosA*	0.003			*fosA*	0.001
Fluoroquinolones	P			CIP		CIP		CIP				CIP	
	G			*oqxA*	0.002	*oqxA*	0.001	*oqxA*	0.001	*oqxA*	0.052	*oqxA*	0.001
				oqxB	0.002	*oqxB*	0.001	*oqxB*	0.001			*oqxB*	0.001
				qrnS1	0.631	*qnrS1*	0.760	*qnrB1*	0.029			*qnrB1*	0.946
								aac(6′)-Ib-cr	0.961			*aac(6′)-Ib-cr*	0.961
QRDR [9]				wild type		wild type		*gyrA* (Ser83-Ile)				*gyrA* (Ser83-Ile)	
QRDR				Wild type		wild type		*parC* (Ser80-Ile)				*parC* (Ser80-Ile)	

[1] ST—sequence type obtained after MLST; [2] n.a.—not applicable; [3] core genome multi-locus sequence-based type; [4] PPP—Posterior probability scores; [5] P—phenotypic resistance: CTX-cefotaxime, CAZ-ceftazidime, FOX-cefoxitin, ATM-aztreonam, GEN-gentamicin, TOB-tobramycin, TET-tetracycline, DOX-doxycycline, CHL-chloramphenicol, SXT-trimethoprim/sulfamethoxazole, CIP-ciprofloxacin; [6] G-genotypic resistance; [7] bold and underline—detected on transconjugats/transformants; [8] bold—within class 1 integron; [9] mutation in *gyrA* and *parC* of quinolone resistance-determining region (QRDR).

Table 2. Identified plasmids in *Klebsiella* isolates.

ID	Plasmid	Identity	Accession Number
1505	IncFIA(HI1)	100.0	AF250878
	IncFIB(pHCM2)	96.49	AL513384
	IncHI1A	99.52	AF250878
	IncHI1B(R27)	100.0	AF250878
	IncQ1	100.0	M28829.1
1635	IncFIA(HI1)	100	AF250878
	IncFIB(pHCM2)	96.49	AL513384
	IncHI1A	99.52	AF250878
	IncHI1B(R27)	100	AF250878
	IncQ1	100	M28829.1
2341b	IncI1	100	AP005147
	IncN	99.61	AY046276
2668	IncN	100	AY046276
	IncR	100	DQ449578
2742	Col440I	92.11	CP023920.1
2826	IncFIA(HI1)	100	AF250878
	IncFIB(pHCM2)	96.49	AL513384
	IncHI1A	99.52	AF250878
	IncHI1B(R27)	100	AF250878
	IncQ1	100	M28829
4545	IncFIB(K)	98.93	JN233704
	Col440I	94.74	CP023920.1

Table 3. Identified virulence factors in four *K. pneumoniae* isolates. Numbers correspond to the exact alleles detected.

Virulence Gene	2341b	2668	2742	4545	
iutA	new allele	new allele	new allele	new allele	aerobactin transport
mrkA	2	2	6	12	
mrkB	33	2	3	2	
mrkC	new allele	2	2	new allele	
mrkD	1	12	12	8	type 3 fimbrial gene cluster
mrkF	new allele	8	8	4	
mrkH	10	7	7	2	
mrkI	7	15	15	4	
mrkJ	19	12	12	2	
ybtA		1			
ybtE		4			
ybtP		4			
ybtQ		22			
ybtS		6			
ybtT		1			yersiniabactin
ybtU		14			
ybtX		15			
fyuA		17			
irp1		44			
irp2		37			

All four *K. pneumoniae* isolates belonged to different sequence types (ST) and cgMLST complex type (CT) (ST147-CT1202, ST307-CT4645, ST1228-CT4644 and a new ST4848-4643). These ST belonged to 4 clonal complexes: CC147 (ST147), CC37 (ST1228), CC307 (ST307) and CC702 (new ST4848). The minimum number of allelic differences between the isolates was 3686.

4. Discussion

The present study demonstrates that broad-spectrum cephalosporin-resistant members of the genus *Klebsiella* are present in the Austrian horse population, although the prevalence in clinical samples seems to be low. These findings are in accordance with previous studies describing the presence of these particular bacteria in horse populations of other countries [4,6–10]. Moreover, a previous study carried out in 2018 on clinical samples from Austrian patients reported 8.4% of *K. pneumoniae* isolates as resistant to third-generation cephalosporins [23].

In the present study, the most prevalent cefotaximase type was CTX-M-1 carried by all three *Klebsiella* species identified; this β-lactamase is commonly associated with *Enterobacteriales* from livestock [24]. CTX-M-15, the dominating cefotaximase, is considered the most common ESBL in *K. pneumoniae* from humans and animals worldwide [4]. To the best of our knowledge, the present study describes for the first time CTX-M-1 producing *K. michiganensis*, a close relative of *K. oxytoca*. One *K. pneumoniae* isolate that displayed both the AmpC and ESBL phenotype carried three different β-lactamases, bla_{CMY2}, bla_{SHV11}, and bla_{TEM-1}. bla_{CMY2} was carried by an IncI1 conjugative plasmid. *K. pneumoniae* carrying plasmid-borne AmpC cephalosporinases (pAmpC) is a rare observation [6,25].

Another important observation is the co-existence of an arsenal of virulence factors and antibiotic resistance characters in one *K. pneumoniae* isolate (ST1228-CT4644). This isolate carried the yersiniabactin locus. Yersiniabactin is a siderophore, which is strongly associated with invasive clinical manifestations in humans [26]. Another siderophore, aerobactin, as well as type 3 fimbriae, which were detected in all *K. pneumoniae* isolates, may enhance colonization and adherence to host cells, invasiveness, and biofilm formation [27].

Among *K. pneumoniae* isolates examined, four different sequence types belonging to four different clonal complexes were identified. Two of these STs, ST147, and ST37, have been recognized as high-risk epidemic multiresistant human-associated clonal lineages [5]. ST147-CC147 is a human-related clone notorious for its multi-drug resistant character and harboring different β-lactamases, including carbapenemases [5]. Recently, this particular clone has emerged in companion animals [5,28]. In contrast, ST1228 has only one entry in the Institut Pasteur MLST database (http://bigsdb.pasteur.fr) and to our knowledge, had never been associated with horses. ST1228 belongs to CC37 whose predicted founder is ST37. *K. pneumoniae* ST37 isolates have been associated with different resistance properties, including carbapenem and colistin resistance, and were isolated from humans and animals [5,29]. One fecal isolate analyzed in the present study belonged to ST307-CC307. ST307 is a relatively new but highly successful pandemic clone, which was previously recovered from human patients, and recent data suggest a multi-drug resistant character of this clone [30]. β-lactamase producing *K. pneumoniae* ST307 has also been detected among different animals [31]. In the present study, a new sequence type, ST4848, belonging to the clonal complex CC702 (predicted founder ST702), has been identified by eBURST analysis. CC702 is a rare clone that has never been associated with broad-spectrum cephalosporin-resistant *K. pneumoniae* of equine origin. Data generated in this study (mating experiments, PlasmidFinder analysis, posterior probability plasmid analysis) strongly suggest that the majority of resistance genes are plasmid-borne. All identified replicons (IncFIA(HI1), IncFIB(K), IncFIB(pHCM2), IncHI1A, IncHI1B(R27), IncI1, IncN, IncQ1, IncR) are considered as vehicles of $bla_{CTX-M-15}$ and $bla_{CTX-M-1}$ dissemination in humans and animals [5,32].

5. Conclusions

Even though the overall prevalence of broad-spectrum cephalosporin-resistant *Klebsiella* sp. among specimens of equine origin in Austria appears to be low, the proportion of broad-spectrum cephalosporin-resistant *Klebsiella* spp. vs. non-resistant *Klebsiella* spp. is worth mentioning, since commensal *Klebsiella* spp. can acquire antimicrobial resistance. As such, the broad-spectrum cephalosporin-resistant *Klebsiella* spp. especially in combination with other resistance properties, are of special clinical importance because of dramatically narrowing the possibility of antibiotic treatment. Due to the regular contact and proximity between horses and humans monitoring horses for the presence of cephalosporin-resistant *Klebsiella* spp. is advisable in order to prevent further spread of these zoonotic agents.

Author Contributions: Conceptualization, I.L., T.L. and J.S.; methodology, I.L., M.P.S. and W.R.; software, M.P.S.; validation, I.L. and M.P.S.; formal analysis, I.L., A.C.R. and W.R.; investigation, I.L., A.C.R., M.P.S. and W.R.; resources, F.A. and J.S.; data curation, I.L., A.C.R. and T.L.; writing—original draft preparation, I.L.; writing—review and editing, I.L., A.C.R., M.P.S., T.L., F.A., W.R. and J.S.; project administration, I.L., F.A., W.R., J.S.; funding acquisition, W.R. All authors have read and agreed to the published version of the manuscript.

Funding: Part of the sequencing-work was funded by a grant awarded under the "MedVetKlebs" Horizon 2020 Framework Programme H2020-SFS-2016-2017 (H2020-SFS-2017-1). Open Access Funding by the University of Veterinary Medicine Vienna.

Acknowledgments: We thank the team of curators of the Institut Pasteur MLST and whole genome MLST databases for curating the data and making them publicly available at http://bigsdb.pasteur.fr. We would also like to express our thanks to Michael Steinbrecher, Anna Stöger, and Barbara Tischler for technical assistance.

Conflicts of Interest: The authors declare no conflict of interest.

References

1. Mulani, M.S.; Kamble, E.E.; Kumkar, S.N.; Tawre, M.S.; Pardesi, K.R. Emerging Strategies to Combat ESKAPE Pathogens in the Era of Antimicrobial Resistance: A Review. *Front. Microbiol.* **2019**, *10*, 539. [CrossRef] [PubMed]

2. Woodford, N.; Turton, J.F.; Livermore, D.M. Multiresistant Gram-negative bacteria: The role of high-risk clones in the dissemination of antibiotic resistance. *FEMS Microbiol Rev.* **2011**, *35*, 736–755. [CrossRef] [PubMed]

3. WHO. Global Priority List of Antibiotic-Resistant Batceria to Guide Research, Discovery, and Development of New Antibiotics. 2017. Available online: http://www.who.int/medicines/publications/WHO-PPL-Short_Summary_25Feb-ET_NM_WHO.pdf (accessed on 1 January 2020).

4. Ewers, C.; Stamm, I.; Pfeifer, Y.; Wieler, L.H.; Kopp, P.A.; Schønning, K.; Prenger-Berninghoff, E.; Scheufen, S.; Stolle, I.; Günther, S.; et al. Clonal Spread of Highly Successful ST15-CTX-M-15 *Klebsiella pneumoniae* in Companion Animals and Horses. *J. Antimicrob. Chemother.* **2014**, *69*, 2676–2680. [CrossRef] [PubMed]

5. Navon-Venezia, S.; Kondratyeva, K.; Carattoli, A. *Klebsiella pneumoniae*: A Major Worldwide Source and Shuttle for Antibiotic Resistance. *FEMS Microbiol. Rev.* **2017**, *41*, 252–275. [CrossRef]

6. Vo, A.T.T.; van Duijkeren, E.; Fluit, A.C.; Gaastra, W. Characteristics of Extended-Spectrum Cephalosporin-Resistant *Escherichia coli* and *Klebsiella pneumoniae* Isolates from Horses. *Vet. Microbiol.* **2007**, *124*, 248–255. [CrossRef]

7. Börjesson, S.; Greko, C.; Myrenås, M.; Landén, A.; Nilsson, O.; Pedersen, K. A Link between the Newly Described Colistin Resistance Gene Mcr-9 and Clinical *Enterobacteriaceae* Isolates Carrying BlaSHV-12 from Horses in Sweden. *J. Glob. Antimicrob. Resist.* **2019**. [CrossRef]

8. Da Roza, F.T.; Couto, N.; Carneiro, C.; Cunha, E.; Rosa, T.; Magalhães, M.; Tavares, L.; Novais, Â.; Peixe, L.; Rossen, J.W.; et al. Commonality of Multidrug-Resistant *Klebsiella pneumoniae* ST348 Isolates in Horses and Humans in Portugal. *Front. Microbiol.* **2019**, *10*, 1657. [CrossRef]

9. Schmiedel, J.; Falgenhauer, L.; Domann, E.; Bauerfeind, R.; Prenger-Berninghoff, E.; Imirzalioglu, C.; Chakraborty, T. Multiresistant Extended-Spectrum β-Lactamase-Producing *Enterobacteriaceae* from Humans, Companion Animals and Horses in Central Hesse, Germany. *BMC Microbiol.* **2014**, *14*, 187. [CrossRef]

10. Shnaiderman-Torban, A.; Paitan, Y.; Arielly, H.; Kondratyeva, K.; Tirosh-Levy, S.; Abells-Sutton, G.; Navon-Venezia, S.; Steinman, A. Extended-Spectrum β-Lactamase-Producing *Enterobacteriaceae* in Hospitalized Neonatal Foals: Prevalence, Risk Factors for Shedding and Association with Infection. *Animals* **2019**, *9*, 600. [CrossRef]

11. Clinical and Laboratory Standards Institute. *Performance Standards for Antimicrobial Susceptibility Testing*, 26th ed.; CLSI supplement M100S; CLSI: Wayne, PA, USA, 2016; pp. 74–80.

12. Lepuschitz, S.; Huhulescu, S.; Hyden, P.; Springer, B.; Rattei, T.; Allerberger, F.; Mach, R.L.; Ruppitsch, W. Characterization of a Community-Acquired-MRSA USA300 Isolate from a River Sample in Austria and Whole Genome Sequence Based Comparison to a Diverse Collection of USA300 Isolates. *Sci. Rep.* **2018**, *8*, 9467. [CrossRef]

13. Bankevich, A.; Nurk, S.; Antipov, D.; Gurevich, A.A.; Dvorkin, M.; Kulikov, A.S.; Lesin, V.M.; Nikolenko, S.I.; Pham, S.; Prjibelski, A.D.; et al. SPAdes: A New Genome Assembly Algorithm and Its Applications to Single-Cell Sequencing. *J. Comput. Biol.* **2012**, *19*, 455–477. [CrossRef] [PubMed]

14. Lepuschitz, S.; Schill, S.; Stoeger, A.; Pekard-Amenitsch, S.; Huhulescu, S.; Inreiter, N.; Hartl, R.; Kerschner, H.; Sorschag, S.; Springer, B.; et al. Whole Genome Sequencing Reveals Resemblance between ESBL-Producing and Carbapenem Resistant *Klebsiella pneumoniae* Isolates from Austrian Rivers and Clinical Isolates from Hospitals. *Sci. Total Environ.* **2019**, *662*, 227–235. [CrossRef] [PubMed]

15. Richter, M.; Rosselló-Móra, R.; Oliver Glöckner, F.; Peplies, J. JSpeciesWS: A Web Server for Prokaryotic Species Circumscription Based on Pairwise Genome Comparison. *Bioinformatics* **2016**, *32*, 929–931. [CrossRef] [PubMed]

16. Alcock, B.P.; Raphenya, A.R.; Lau, T.T.Y.; Tsang, K.K.; Bouchard, M.; Edalatmand, A.; Huynh, W.; Nguyen, A.-L.V.; Cheng, A.A.; Liu, S.; et al. CARD 2020: Antibiotic Resistome Surveillance with the Comprehensive Antibiotic Resistance Database. *Nucleic Acids Res.* **2019**, *48*, D517–D525. [CrossRef] [PubMed]

17. Zankari, E.; Hasman, H.; Cosentino, S.; Vestergaard, M.; Rasmussen, S.; Lund, O.; Aarestrup, F.M.; Larsen, M.V. Identification of Acquired Antimicrobial Resistance Genes. *J. Antimicrob. Chemother.* **2012**, *67*, 2640–2644. [CrossRef]

18. Carattoli, A.; Zankari, E.; Garcìa-Fernandez, A.; Larsen, M.; Lund, O.; Villa, L.; Aarestrup, F.; Hasman, H. PlasmidFinder and PMLST: In Silico Detection and Typing of Plasmids. *Antimicrob. Agents Chemother.* **2014**, *58*, 3895–3903. [CrossRef]

19. Arredondo-Alonso, S.; Rogers, M.R.C.; Braat, J.C.; Verschuuren, T.D.; Top, J.; Corander, J.; Willems, R.J.L.; Schürch, A.C. Mlplasmids: A User-Friendly Tool to Predict Plasmid- and Chromosome-Derived Sequences for Single Species. *Microb. Genom.* **2018**, *4*. [CrossRef]

20. Desvars-Larrive, A.; Ruppitsch, W.; Lepuschitz, S.; Szostak, M.P.; Spergser, J.; Feßler, A.T.; Schwarz, S.; Monecke, S.; Ehricht, R.; Walzer, C.; et al. Urban Brown Rats (*Rattus Norvegicus*) as Possible Source of Multidrug-Resistant *Enterobacteriaceae* and Meticillin-Resistant *Staphylococcus* Spp., Vienna, Austria, 2016 and 2017. *Eurosurveillance* **2019**, *24*. [CrossRef]

21. Loncaric, I.; Beiglböck, C.; Feßler, A.T.; Posautz, A.; Rosengarten, R.; Walzer, C.; Ehricht, R.; Monecke, S.; Schwarz, S.; Spergser, J.; et al. Characterization of ESBL- and AmpC-Producing and Fluoroquinolone-Resistant *Enterobacteriaceae* Isolated from Mouflons (*Ovis orientalis musimon*) in Austria and Germany. *PLoS ONE* **2016**, *11*, e0155786. [CrossRef]

22. Sweeney, M.T.; Lubbers, B.V.; Schwarz, S.; Watts, J.L. Applying Definitions for Multidrug Resistance, Extensive Drug Resistance and Pandrug Resistance to Clinically Significant Livestock and Companion Animal Bacterial Pathogens. *J. Antimicrob. Chemother.* **2018**, *73*, 1460–1463. [CrossRef]

23. European Centre for Disease Prevention and Control. *Surveillance of Antimicrobial Resistance in Europe Annual Report of the European Antimicrobial Resistance Surveillance Network (EARS-Net) 2018*; European Centre for Disease Prevention and Control: Solna kommun, Sweden, 2019. [CrossRef]

24. Ewers, C.; Bethe, A.; Semmler, T.; Guenther, S.; Wieler, L.H. Extended-Spectrum β-Lactamase-Producing and AmpC-Producing *Escherichia coli* from Livestock and Companion Animals, and Their Putative Impact on Public Health: A Global Perspective. *Clin. Microbiol. Infect.* **2012**, *18*, 646–655. [CrossRef] [PubMed]

25. Madec, J.Y.; Haenni, M.; Nordmann, P.; Poirel, L. Extended-spectrum β-lactamase/AmpC- and carbapenemase-producing *Enterobacteriaceae* in animals: A threat for humans? *Clin. Microbiol. Infect.* **2017**, *23*, 826–833. [CrossRef] [PubMed]

26. Holt, K.E.; Wertheim, H.; Zadoks, R.N.; Baker, S.; Whitehouse, C.A.; Dance, D.; Jenney, A.; Connor, T.R.; Hsu, L.Y.; Severin, J.; et al. Genomic Analysis of Diversity, Population Structure, Virulence, and Antimicrobial Resistance in *Klebsiella pneumoniae*, an Urgent Threat to Public Health. *Proc. Natl. Acad. Sci. USA* **2015**, *112*, E3574–E3581. [CrossRef] [PubMed]

27. Clegg, S.; Murphy, C.N. Epidemiology and Virulence of *Klebsiella pneumoniae*. *Microbiol. Spectr.* **2016**, *4*. [CrossRef]

28. Ovejero, C.M.; Escudero, J.A.; Thomas-Lopez, D.; Hoefer, A.; Moyano, G.; Montero, N.; Martin-Espada, C.; Gonzalez-Zorn, B. Highly Tigecycline-Resistant *Klebsiella pneumoniae* Sequence TYPE 11 (ST11) & ST147 Isolates from Companion Animals. *Antimicrob. Agents Chemother.* **2017**, *61*, e02640–e02716. [CrossRef]

29. Taniguchi, Y.; Maeyama, Y.; Ohsaki, Y.; Hayashi, W.; Osaka, S.; Koide, S.; Tamai, K.; Nagano, Y.; Arakawa, Y.; Nagano, N. Co-Resistance to Colistin and Tigecycline by Disrupting MgrB and RamR with IS Insertions in a Canine *Klebsiella pneumoniae* ST37 Isolate Producing SHV-12, DHA-1 and FosA3. *Int. J. Antimicrob. Agents* **2017**, *50*, 697–698. [CrossRef]

30. Wyres, K.L.; Hawkey, J.; Hetland, M.A.K.; Fostervold, A.; Wick, R.R.; Judd, L.M.; Hamidian, M.; Howden, B.P.; Löhr, I.H.; Holt, K.E. Emergence and Rapid Global Dissemination of CTX-M-15-Associated *Klebsiella pneumoniae* Strain ST307. *J. Antimicrob. Chemother.* **2019**, *74*, 577–581. [CrossRef]

31. Harada, K.; Shimizu, T.; Mukai, Y.; Kuwajima, K.; Sato, T.; Usui, M.; Tamura, Y.; Kimura, Y.; Miyamoto, T.; Tsuyuki, Y.; et al. Phenotypic and Molecular Characterization of Antimicrobial Resistance in *Klebsiella* Spp. Isolates from Companion Animals in Japan: Clonal Dissemination of Multidrug-Resistant Extended-Spectrum β-Lactamase-Producing *Klebsiella pneumoniae*. *Front. Microbiol.* **2016**, *7*, 1021. [CrossRef]

32. Carattoli, A. Plasmids and the Spread of Resistance. *Int. J. Med. Microbiol.* **2013**, *303*, 298–304. [CrossRef]

Article

Extended-Spectrum β-lactamase-Producing *Enterobacteriaceae* Shedding in Farm Horses Versus Hospitalized Horses: Prevalence and Risk Factors

Anat Shnaiderman-Torban [1], Shiri Navon-Venezia [2,3,†], Ziv Dor [2], Yossi Paitan [4,5], Haia Arielly [5], Wiessam Abu Ahmad [6], Gal Kelmer [1], Marcus Fulde [7] and Amir Steinman [1,*,†]

[1] Koret School of Veterinary Medicine (KSVM), The Robert H. Smith Faculty of Agriculture, Food and Environment, The Hebrew University of Jerusalem, Rehovot 7610001, Israel; ashnaiderman@gmail.com (A.S.-T.); gal.kelmer@mail.huji.ac.il (G.K.)

[2] Department of Molecular Biology, Faculty of Natural Sciences, Ariel University, Ariel 40700, Israel; shirinv@ariel.ac.il (S.N.-V.); zivddor@gmail.com (Z.D.)

[3] The Miriam and Sheldon Adelson School of Medicine, Ariel University, Ariel 40700, Israel

[4] Department of Clinical Microbiology and Immunology, Sackler Faculty of Medicine, Tel Aviv University, Tel Aviv 6997801, Israel; yossi.paitan@clalit.org.il

[5] Clinical Microbiology Lab, Meir Medical Center, Kfar Saba 4428164, Israel; Ariellyhaya@clalit.org.il

[6] Braun School of Public Health and Community Medicine, Hebrew University, Jerusalem 9112102, Israel; wiessam@gmail.com

[7] Institute of Microbiology and Epizootics, Department of Veterinary Medicine at the Freie Universität Berlin, Berlin 14163 Germany; Marcus.Fulde@fu-berlin.de

* Correspondence: amirst@savion.huji.ac.il

† These authors had equal contribution.

Received: 15 January 2020; Accepted: 7 February 2020; Published: 11 February 2020

Simple Summary: This prospective study investigated the prevalence, molecular characteristics and risk factors of extended-spectrum β-lactamase (ESBL)-producing Enterobacteriaceae (ESBL-E) shedding in three equine cohorts: (i) farm horses (13 farms, n = 192); (ii) on admission to a hospital (n = 168) and; (iii) horses hospitalized for ≥72 h re-sampled from cohort (ii) (n = 86). Bacteria were isolated from rectal swabs, identified, antibiotic susceptibility patterns were determined, and medical records and owners' questionnaires were analyzed for risk factor analysis. ESBL shedding rates significantly increased during hospitalization (77.9%, n = 67/86), compared to farms (20.8%, n = 40/192), and horses on admission (19.6%, n = 33/168). High bacterial species diversity was identified, mainly in cohorts (ii) and (iii), with high resistance rates to commonly used antimicrobials. Risk factors for shedding in farms included horses' breed (Arabian), sex (stallion), and antibiotic treatment. Older age was identified as a protective factor. We demonstrated a reservoir for antibiotic-resistant bacteria in an equine hospital and farms, with a significant ESBL-E acquisition. In light of our findings, in order to control ESBL spread, we recommend conducting active ESBL surveillance programs alongside antibiotic stewardship programs in equine facilities.

Abstract: We aimed to investigate the prevalence, molecular characteristics and risk factors of extended-spectrum β-lactamase (ESBL)-producing *Enterobacteriaceae* (ESBL-E) shedding in horses. A prospective study included three cohorts: (i) farm horses (13 farms, n = 192); (ii) on hospital admission (n = 168) and; (iii) horses hospitalized for ≥72 h re-sampled from cohort (ii) (n = 86). Enriched rectal swabs were plated, ESBL-production was confirmed (Clinical and Laboratory Standards Institute (CLSI)) and genes were identified (polymerase chain reaction (PCR)). Identification and antibiotic susceptibility were determined (Vitek-2). Medical records and owners' questionnaires were analyzed. Shedding rates increased from 19.6% (n = 33/168) on admission to 77.9% (n = 67/86) during hospitalization ($p < 0.0001$, odds ratio (OR) = 12.12). Shedding rate in farms was 20.8% (n = 40/192), significantly lower compared to hospitalized horses ($p < 0.0001$). The main ESBL-E species (n = 192 isolates) were *E. coli* (59.9%, 115/192), *Enterobacter* sp. (17.7%, 34/192) and *Klebsiella pneumoniae* (13.0%, 25/192). The main gene group was CTX-M-1 (56.8%). A significant increase in resistance rates to chloramphenicol, enrofloxacin, gentamicin, nitrofurantoin, and trimethoprim-sulpha was identified during hospitalization. Risk factors for shedding in farms included breed (Arabian, OR = 3.9), sex (stallion, OR = 3.4), and antibiotic treatment (OR = 9.8). Older age was identified as a protective factor (OR = 0.88). We demonstrated an ESBL-E reservoir in equine cohorts, with a significant ESBL-E acquisition, which increases the necessity to implement active surveillance and antibiotic stewardship programs.

Keywords: equine; ESBL-E; antibiotic resistance; shedding; risk factors; farm; ESBL-E acquisition

---

1. Introduction

Extended-spectrum beta-lactamase (ESBL)-producing *Enterobacteriaceae* (ESBL-E) poses a clinical challenge to both human and veterinary clinicians. ESBLs confer resistance to penicillins, cephalosporins, and aztreonam and are often accompanied by fluoroquinolone resistance, which even further narrows antibiotic treatment options [1]. Moreover, many ESBL genes are encoded on large plasmids, which enables lateral transfer between different bacterial species, within the same host and between different hosts [2]. In human medicine, ESBL production is associated with increased morbidity, higher overall and infection-related mortality, increased hospital length of stay, delay of targeted appropriate treatment, and higher costs [3,4]. Risk factors for colonization and infection in humans include severe illness with prolonged hospital stays, the presence of invasive medical devices for a prolonged duration and antibiotic use [2].

Within the last decade, a growing burden of ESBL-E in companion animals is being observed, both as gut colonizing bacteria and as infecting pathogens, causing wounds, respiratory, urogenital, gastro-intestinal, umbilical infections, and bacteremia [5–8]. Horses were described as carriers, as well as infected by ESBL-E, in equine clinics and in farm settings [9,10]. Prevalence of ESBL-producing *E. coli* carriage in horses varies between 4–44% in different European countries [11–13], with a lower carriage prevalence in equine riding centers in comparison with equine clinics [10]. In equine community settings, being stabled in the same yard with a recently hospitalized horse was identified as a risk factor for ESBL-producing *E. coli* carriage [14]. Risk factor analysis in the level of the farm revealed that the odds of being an ESBL/AmpC-producing *E. coli* premises were higher among riding schools than breeding premises, if premises housed a horse that had been medically treated with antibiotics within the last three months, and also in premises where the staff consisted of more than five persons [13]. However, risk factors for shedding of different ESBL-E species within horses were not yet reported.

We aimed to investigate and compare ESBL-E shedding in different equine cohorts, including farm horses, horses on admission to an equine hospital and during hospitalization, as well as to determine risk factors for shedding. We hypothesized that shedding rates increase during hospitalization, that previous antibiotic treatment is a risk factor for shedding and that shedding on admission and during hospitalization is associated with clinical signs, prolonged hospitalization, and severe outcome.

2. Materials and Methods

2.1. Equine Study Cohorts, Study Design, and Sampling Methods

This prospective study was performed on 13 farms throughout Israel and in the Koret School of Veterinary Medicine—Veterinary Teaching Hospital (KSVM-VTH). The study was approved by the Internal Research Review Committee of the KSVM-VTH (Reference numbers: KSVM-VTH/15_2015, KSVM-VTH/23_2015). Rectal swabs were collected from the horses with owner consent. On admission, sampling was performed prior to any medical treatment in the hospital. When horses survived and were not discharged, a second sample was taken 72 h post-admission. Farm horses were located in different regions of Israel to roughly represent the population.

2.2. Demographic and Medical Data

For farm horses (cohort (i)), owners' questionnaires were reviewed for data regarding individual horses, including the originating farm, signalment (age, sex, and breed), duration of the horse's accommodation in the farm, hospitalization and antibiotic treatments within the previous year.

For hospitalized horses (cohort (ii)), medical records were reviewed for the following information: signalment (age, sex, and breed), geographic origin, previous admission to the hospital within the previous year (yes/no), clinical signs, duration of illness before admission, antibiotic therapy before and during hospitalization, surgical procedures, other medications, hospitalization length, short-term outcome, and admission charge.

2.3. ESBL-E Isolation and Species Identification

Rectal specimens [14] were collected using bacteriological swabs (Meus s.r.l., Piove di Sacco, Italy) and were inoculated directly into a Luria Bertani infusion enrichment broth (Hy-Labs, Rehovot, Israel) to increase the sensitivity of ESBL-E detection [15]. After incubation at 37 °C (18–24 h), enriched samples were plated onto Chromagar ESBL plates (Hy-Labs, Rehovot, Israel), at 37 °C for 24 h. Colonies that appeared after overnight incubation at 37 °C were recorded, and one colony of each distinct color was re-streaked onto a fresh Chromagar ESBL plate to obtain a pure culture. Pure isolates were stored at −80 °C for further analysis.

Isolates were subjected to Vitek-MS (BioMérieux, Inc., Marcy-l'Etoile, France) for species identification or to Vitek-2 (BioMérieux, Inc., Marcy-l'Etoile, France) for species identification and/or antibiotic susceptibility testing (AST-N270 Vitek 2 card). Chloramphenicol, enrofloxacin, and imipenem were analyzed using disc diffusion assay (Oxoid, Basingstoke, UK). ESBL-production was confirmed by combination disk diffusion using cefotaxime and ceftazidime discs (Oxoid, Basingstoke, UK), as well as cefotaxime and ceftazidime with clavulanic acid (Sensi-Discs BD, Breda, The Netherlands). Results were interpreted according to the Clinical and Laboratory Standards Institute (CLSI) guidelines [16]. Multidrug-resistant (MDR) bacteria were defined as such due to their in vitro resistance to three or more classes of antimicrobial agents [17].

2.4. Molecular Characterization of ESBL-E

Isolates were examined for the presence of the *bla*CTX-M group using a multiplex polymerase chain reaction (PCR) from ESBL-E DNA lysates, as previously described [18]. Isolates that were found to be *bla*CTX-M PCR negative were further examined for the presence of *bla*OXA-1, *bla*OXA2, *bla*OXA10 [19], *bla*TEM, and *bla*SHV groups [20]. ESBL-producing *E. coli* isolates were subjected to PCR for the detection of *mdh* and *gyrB* genes in order to determine the presence of the worldwide pandemic *E. coli* ST131 lineage [21].

2.5. Sample Size and Statistical Analysis

The minimal sample size (number of animals sampled) for farm horses was calculated using WinPepi, based on an estimated shedding rate of 25% for ESBL-E in equine community livery premises [22] and on the fact that Israel is endemic for ESBL-E [23], with a confidence level of 95% and an acceptable difference of 7%, resulting in n = 147.

The minimal sample size for horses on admission to hospital was based on the expected difference between ESBL-E shedding and non-shedding horses and the percentage of admitted horses that were treated with antibiotics before admission since antibiotic treatment was assumed to be a risk factor for shedding [12]. Since there is no previous study revealing percentages of antibiotic-treated horses and ESBL shedding, data for this calculation was based on a human study [24]. Estimating that 25% of horses on admission are ESBL shedders (representing the equine community) and that 72%

and 44% of horses were treated with antimicrobials within shedders and non-shedders, respectively, with a 5% significance level and power of 80%, the total required sample size is 145 horses, including 116 non-shedders and 29 shedders.

Risk assessment was performed using Chi-square or Fisher's exact tests for association between individual variables, shedding and ESBL-E acquisition. Descriptive statistics were used to describe shedding rates. Continuous variables were analyzed using t-tests or Mann–Whitney U-tests. $p \leq 0.05$ was considered statistically significant. For risk factor analysis of farm horses, a logistic regression model (multivariable analysis) was conducted using all the significant variables in the univariable analysis at a significance level of $p < 0.2$ using the ENTER method (IBM SPSS Statistics 25). Categorical data were summarized by the number of cases (percentage) and confidence intervals (95%) were calculated by Fisher's (WinPEPI 11.15 Describe A).

In order to compare between shedding rates and antibiotic resistance rates within horses on admission and during hospitalization (cohorts (ii) and (iii), respectively), a mixed effect logistic regression model was conducted (STATA version 13). Resistance was defined as complete resistance (not including "intermediate resistance"). Odds ratio (OR) for a significant change in antibiotic resistance rates is defined as OR for a change in one resistance category (e.g., a change from "susceptible" to "intermediate" or from "intermediate" to "resistant"). A comparison between shedding rates and antibiotic resistance rates between farm horses (cohort (i)) and horses on admission (cohort (ii)) was performed using Chi-square.

3. Results

3.1. Characterization of the Equine Study Populations (Table 1)

Overall, 192 horses were sampled, originating from 13 farms across Israel (June 2016–September 2018). The average number of sampled horses per farm was 15 (range: 3–26 horses).

On admission, 168 horses were sampled (November 2015 to April 2016). Horses were admitted to hospitalization due to the following reasons: gastro-intestinal pathologies (33%, n = 55/168), orthopedic disorders (17%, n = 29/168), healthy (mares of sick neonatal foals or foals of sick mares, 17%, n = 29/168), reproduction disorders (12%, n = 20/168), neonatology disorders (12%, n = 20/168), respiratory disorders (4%, n = 7/168), and others (including ophthalmic, hematology, endocrine, teeth disorders, and tumors, 5%, n = 8/168). The median length of illness before admission was one day (range: several hours–750 d). Horses hospitalized for ≥72 h were re-sampled (n = 86).

Table 1. Characterization of farm horses versus horses on admission to hospital.

Equine Cohort	Breeds [1]	Median Age [2] (Years ± SD)	Sex Distribution [3]
Farm horses (n = 192)	41.1% Arabians (n = 79/192) 25% pacers (n = 48/192) 15.1% Quarter horses (n = 29/192) 9.9% Warmbloods (n = 19/192) 5.2% local breed (n = 10/192) 3.7% ponies (n = 7/192)	8 ± 5.3	mares (72.4%, n = 139/192) geldings (12.5%, n = 24/192) stallions (11.5%, n = 22/192) [4]
Horses on admission (n = 168)	49.4% Arabians (n = 83/168) 19.6% Quarter horses (n = 33/168) 14.3% pacers (n = 24/168) 7.7% Friesians (n = 13/168) 4.8% Warmbloods (n = 8/168) 4.2% others (n = 7/168)	4.5 ± 5.2	mares (68.5%, n = 115/168) geldings (16.1%, n = 27/168) stallions (15.4%, n = 26/168)

[1] Breed distribution was not significantly different for Arabians, Quarter horses, and Warmbloods in comparison to farm horses, and was significantly different for the pacers horses (significantly higher in farms, $p = 0.012$) and Friesians (significantly higher on admission, $p < 0.001$); [2] Median age of horses on admission was significantly lower than the median age of farm horses ($p < 0.0001$); [3] Sex distribution was not significantly different between farm horses and horses on admission; [4] Data was not available for seven horses.

3.2. Antibiotic Therapy, Surgical Procedures, Length of Stay, and Outcome

A proportion of 8.3% (n = 16/192) of farm horses was hospitalized within the previous year, ranging from 0–30% between farms. A proportion of 19.8% (n = 38/192) of horses were treated with antibiotics within the previous year, ranging from 0–61% between farms. On admission, 9.5% (n = 16/168) of horses were reported to be previously hospitalized (within a year period), and 16.1% (n = 27/168) of horses were treated with antibiotics within the previous year. Previous hospitalization and antibiotic treatment prevalence rates were not significantly different in comparison with farm horses.

During hospitalization, 50.6% of horses (n = 85/168) were treated with antibiotics, a proportion which is significantly higher than antibiotic treatment in farms and prior to admission ($p < 0.0001$). Surgical procedures were performed in 36.9% of horses (n = 62/168). The median length of stay was three days (range: several hours-21 d). Out of all horses admitted to hospitalization, 84.4% survived to discharge (n = 142/168).

3.3. Prevalence of ESBL-E Shedding

Within farm horses, shedding rate was 20.8% [n = 40/192, 95% Confidence interval (CI) 15.3–27.3%, Table 1]. Shedding rate on admission was 19.6% (n = 33/168, 95% CI: 13.9–26.5%), which was not statistically different from shedding rate in farms ($p = 0.79$). Shedding rate of hospitalized horses (re-sampled) was 77.9% (n = 67/86, 95% CI 67.7–86.1%), which was significantly higher than the shedding rate on admission and in farms (p<0.001, OR = 12.12, 95% CI 3.92–37.49). Out of 67 hospitalized shedding horses, 77.6% (n = 52/67, 95% CI 65.8–86.9%) did not shed ESBL-E on admission.

3.4. Distribution of ESBL-E Species and ESBL Genes

Overall, 192 ESBL-E isolates were analyzed (Table A1). Fourteen bacterial species were identified of which three were identified in all cohorts—*E. coli*, *Klebsiella pneumoniae*, and *Enterobacter cloacae* (Figure 1). The most prevalent bacterial species in all cohorts was *E. coli*, consisting of 79.2% of isolates from farms, 66.7% from horses on admission, and 49.0% from hospitalized horses. However, the prevalence of *E. coli* decreased in horses on admission and in hospitalized horses, as the diversity of other ESBL-E species increased, from four species in farms to five species on admission and twelve species in hospitalized horses. Nosocomial ESBL-E species that were not identified in farms and on admission included *Citrobacter freundii* (n = 3/105), *Salmonella* spp (n = 3/105), *K. oxytoca*, *Citrobacter brakii*, *E. vulneris*, *Pantoea* spp, *Proteus mirabilis*, and *Raoultella ornithinolytica* (n = 1/105 each). The pandemic hypervirulent *E. coli* ST131 [25] was identified in three horses: two horses on admission and one horse during hospitalization. The main ESBL gene was the blaCTX-M-1 group in all cohorts (total 56.8% of all isolates, Table 2).

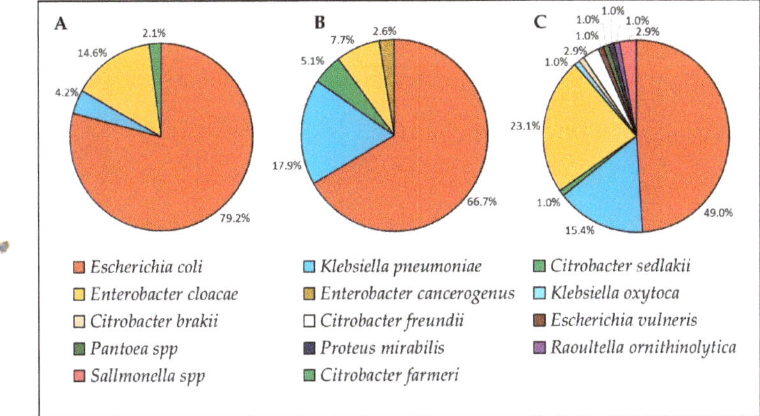

Figure 1. ESBL-E species distribution isolated from cohort (i) farm horses ((**A**), n = 48 isolates), cohort (ii) horses on admission to the hospital ((**B**), n = 39 isolates) and cohort (iii) 72 h post-admission ((**C**), n = 105 isolates).

Table 2. Shedding rates of extended-spectrum beta-lactamase-producing *Enterobacteriaceae* (ESBL-E) in farm horses, on admission, and during hospitalization.

Equine Cohort	Shedding (%)	Total No. of ESBL-E Isolates	MDR Isolates (%)	*bla*ESBL Gene Group (%)
Farm horses	40/192 (20.8) (95% CI: 15.3–27.3%)	48	43/48 (89.6) (95% CI: 77.3–96.5)	CTX-M-1: 35/48 (72.9) CTX-M-9: 1/48 (2.1) CTX-M-25: 1/48 (2.1) SHV-12: 5/48 (10.4)
Horses on admission	33/168 (19.6) (95% CI: 13.9–26.5%)	39	28/39 (71.8) (95% CI: 55.1–85.0%)	CTX-M-1: 24/39 (61.5) CTX-M-9: 1/39 (2.5) SHV-12: 3/39 (7.7) SHV-2: 1/39 (2.5) SHV-28: 1/39 (2.5)
Hospitalized horses (72 h post admission) [1]	67/86 (77.9) [2] (95% CI 67.7–86.1%)	105	99/105 (94.3) (95% CI: 87.9–97.9%) [3]	CTX-M-1: 50/105 (47.6) CTX-M-2: 8/105 (7.6) CTX-M-9: 7/105 (6.7) CTX-M-25: 1/105 (0.95) OXA-1: 2/105 (1.9) SHV-12: 26/105 (24.7) SHV-228: 1/105 (0.95)

[1] Horses re-sampled from cohort "horses on admission"; [2] Shedding rate in hospitalized horses is significantly higher than shedding rate on admission and in farms ($p < 0.0001$, OR=12.12, 95% CI 3.92–37.49); [3] Prevalence of multidrug-resistant (MDR) isolates is significantly higher in isolates originated from hospitalized horses compared to isolates originated from horses on admission ($p < 0.001$).

3.5. Antibiotic Susceptibility Profiles

Antibiotic resistance rates varied between cohorts, with a significant increase during hospitalization. All isolates from all cohorts were susceptible to imipenem (Table 3).

Table 3. Antibiotic [1] resistance rates (percentage) of ESBL-E isolates shed by farm horses, horses on admission, and hospitalized horses.

Equine Cohort	AMP	AMC [2]	LEX	CAZ	IMP	CHL [3]	ENR [4]	AMK	GEN [5]	NIT [6]	TMS [7]
Farms	100	41.7	100	100	0	66.6	6.3	0	75	4.2	89.6
On admission	100	82.1	100	85.0	0	46.2	17.9	2.6	48.7	5.3	76.3
During hospitalization	96.0	32.0	99.0	90.0	0	85.3	51.5	10.8	84.3	11.0	95.0

[1] Abbreviations: ampicillin (AMP), amoxicillin-clavulanate (AMC), cephalexin (LEX), ceftazidime (CAZ), imipenem (IMP), chloramphenicol (CHL), enrofloxacin (ENR), amikacin (AMK), gentamicin (GEN), nitrofurantoin (NIT), and Trimethoprim- sulpha (TMS); [2] An increase in resistance rates for AMC on admission compared to farms ($p = 0.001$) and a decrease during hospitalization compared to admission ($p < 0.001$, OR = 0.1, 95% CI 0.04, 0.26); [3] An increase in resistance rates for CHL during hospitalization compared to admission ($p < 0.001$, OR = 6.5, 95% CI 2.8, 15); [4] An increase in resistance rates for ENR during hospitalization compared to admission ($p < 0.001$, OR = 4.2, 95% CI 1.9, 9.5); [5] An increase in resistance rates for GEN during hospitalization compared to admission ($p < 0.001$, OR = 12.3, 95% CI 2.9, 52.5); [6] An increase in resistance rates for NIT during hospitalization compared to admission ($p < 0.001$, OR = 3.7, 95% CI 1.4, 9.5); [7] An increase in resistance rates for TMS during hospitalization compared to admission ($p < 0.01$, OR = 6, 95% CI 1.9, 19.4).

Among bacteria that grew on Chromagar ESBL plates, the prevalence of MDR bacteria was 89.6%, 71.8%, and 94.3% in farms, horses on admission, and hospitalized horses, respectively. The prevalence rate was significantly higher in isolates originated from hospitalized horses compared to horses on admission ($p = 0.001$, Table 2).

3.6. Risk Factor Analysis for ESBL-E Shedding

3.6.1. Farm Horses

In univariable analysis, horses' breed, sex, hospitalization in the previous year, antibiotic treatment in the previous year, and age were significantly associated with ESBL-E shedding (Table 2). Since the Arabian breed was the most prevalent breed sampled, we clustered all other breeds as one category in the multivariable analysis. In a logistic regression model, the breed (Arabian), sex (stallion versus mare, which was the reference in this category), and antibiotic treatment in the previous year were identified as risk factors for shedding. Age greater than one year was identified as a protective factor (Table 4).

Table 4. Risk factor analysis for ESBL-E shedding by farm horses (logistic regression model).

Variable	*p*-value	Odds Ratio (95% CI)
Breed (Arabian versus non-Arabian)	0.006	3.9 (1.5–10.4)
Sex (reference: mare)	0.079	-
Stallion	0.029	3.4 (1.1–12.2)
Gelding	0.744	0.7 (0.07–6.4)
Age	0.008	0.9 (0.8–0.97)
Hospitalization within the previous year	0.194	2.9 (0.6–14.8)
Antibiotic treatment within the previous year	<0.0001	9.8 (3.6–26.8)

3.6.2. Horses on Admission

Signalment (age, sex, and breed), geographic origin, prior hospitalizations in the last year, clinical signs, length of illness before admission, antibiotic therapy before and during hospitalization, surgical procedures, other medications, hospitalization length, short-term outcome, and admission charge were not associated with ESBL-E shedding on admission (Table 2). Sex, hospitalization length, and admission charge resulted in $p < 0.2$, therefore, were analyzed via a logistic regression model, which did not yield any significant associations (Table 3).

3.6.3. Horses During Hospitalization

There was no association between ESBL shedding 72 h post-admission and on admission, clinical signs on admission, antibiotic treatment during hospitalization, surgical procedures during hospitalization, length of stay, admission charge and outcome (Table 2).

4. Discussion

This study investigates ESBL-E shedding in three equine cohorts, including farm horses, representing community equine, as well as horses on admission to the hospital and during hospitalization. Studies regarding antibiotic-resistant pathogens shedding, either in farm horses or in hospitalized horses were reported previously from different European countries [13,22,26,27]. Our study compares different equine cohorts within the same country. Both community and hospital cohorts are of great interest, from a veterinary and a 'one health' perspective, therefore it is highly valuable to compare these cohorts.

We found high ESBL-E shedding rates (Table 2), an increased bacterial species diversity (Figure 1) as well as in the ESBL-E genes variety (Table 2). An increase in shedding rates may be due to the acquisition of bacteria, plasmids or resistance genes. The main bacterial species in all cohorts was *E. coli*, with decreased incidence on admission and during hospitalization, due to increased incidence of other nosocomial ESBL-producing bacterial species. The main ESBL gene group was CTX-M-1, as was previously reported in community horses [26]. However, on admission and during hospitalization, CTX-M-1 incidence decreases, alongside an increase in the number of ESBL genes. A study conducted in an equine hospital in the UK demonstrated the emergence of ESBL-producing *E. coli* during a decade [26], whereas we demonstrated a significant increase in ESBL-E shedding during individual horses' hospitalization. These findings support an urgent necessity in active surveillance and infection control programs in veterinary facilities and hospitals.

In addition, there is a need to set strict antibiotic stewardship programs in veterinary medicine, specifically in companion animals' facilities, with specific guidance and enforcement. According to a recommendation published by the Committee for Medicinal Products for Veterinary Use (CVMP) of the European Union, there is a need to reserve fluoroquinolones, third and fourth generation cephalosporins for treatment when other options are likely to fail, and whenever possible, treatment should be supported by an antimicrobial susceptibility testing [28]. In practice, fluoroquinolones and cephalosporins are in use in equine medicine, sometimes as a first-line choice [29,30]. In our study, ESBL-E shedding as well as resistance rates for chloramphenicol, enrofloxacin, gentamicin, nitrofurantoin, and trimethoprim-sulpha increased significantly during hospitalization, resulting in a significant increase in MDR bacterial species shedding (Tables 2 and 3). In light of our findings, as well as increasing resistance rates in other equine studies, we recommend implementing antibiotic stewardships in equine clinics and hospitals [31,32].

We also aimed to determine risk factors for shedding. We did not find significant associations between shedding on admission and during hospitalization to medical data. During the study period, we sampled all horses on admission, which represented a heterogeneous population, including critically ill horses alongside healthy mares, which were hospitalized together with their sick foals. Therefore, the lack of significant risk factors may be due to high variation in the equine population. Many of the pathologies on admission were attributed to the gastro-intestinal system, which might influence the intestinal microbiome. However, clinical signs on admission and during hospitalization were not associated with shedding. In farm horses, we detected several risk factors for ESBL-E shedding (Table 4). The Arabian breed was the main breed within farm horses and horses on admission to hospital. These horses in Israel are used mainly for breeding and shows and are held under intensive management, which may explain the risk for ESBL shedding. Interestingly, we detected the 'stallion' sex as a risk factor. In human medicine, it is reported that males are more susceptible to diverse bacterial illnesses than females, including an ESBL-E infection [33], presumably related to hormonal influences [34]. This may explain also our findings in veterinary medicine, however, it requires further investigation. Previous antibiotic treatment was identified as a risk factor as well, in agreement with other human and veterinary studies [2,13]. Age older than one year was identified as a protective factor, which may be due to the maturation of immunity. In a national survey of cattle farms in Israel, the prevalence of ESBL-E was higher in calves versus adult cows, where the use of antimicrobial prophylaxis was more common [35]. In human medicine, elderly age is associated with ESBL-E infections [33]. However, in our study, elderly horses older than 20 years old [36] were not prevalent and consisted of 3% (n = 12/360) of the study population. Therefore, elderly age may not be identified as a risk factor.

Our results should also be addressed from a 'one health' perspective. We detected resistant zoonotic bacteria both in farms and in hospital settings, which underlines the necessity for awareness and improved management. The human-animal interaction has great psychological and physical established benefits, with a great emphasis on equine-assisted therapy [37–39]. Therefore, there is pronounced importance in establishing safety policies involving therapists, physicians, and veterinarians, in order to ensure safe human-equine interactions in community settings [40]. This also applies to veterinary hospital staff. In a longitudinal study involving veterinary hospital staff and students, a higher level of ESBL-producing *E. coli* carriage was observed longitudinally [41], which underlines the necessity to implement gold standards biosecurity programs in veterinary hospitals.

5. Conclusions

Multi-drug resistant potentially zoonotic bacteria were detected both in farm horses and in hospitalized horses, with a significantly increased shedding during hospitalization. Therefore, we recommend implementing active surveillance programs alongside with infection control and antibiotic stewardship policies, in order to decrease resistance burden and to allow safe human-equine interactions.

Author Contributions: Conceptualization, A.S.-T., S.N.-V. and A.S.; methodology, S.N.-V. A.S., W.A.A, Y.P. and H.A.; software, W.A.A.; validation, A.S.-T., S.N.-V. and A.S.; formal analysis, A.S.-T., Z.D., Y.P. and H.A.; investigation, A.S.-T.; resources, S.N.-V., A.S., M.F., Y.P., G.K.; data curation, A.S.-T.; writing—original draft preparation, A.S.-T., S.N.-V. and A.S.; writing—review and editing, all authors; visualization, all authors; supervision, S.N.-V. and A.S.; project administration, A.S.-T., S.N.-V. and A.S.; funding acquisition, S.N.-V. and A.S. All authors have read and agreed to the published version of the manuscript.

Funding: This research received no external funding.

Acknowledgments: We are grateful to the KSVM-VTH equine department staff, farm owners, employees, and veterinarians for their collaboration in conducting this study.

Conflicts of Interest: The authors declare no conflict of interest.

Appendix A

Table A1. Antimicrobial susceptibility profiles of individual isolates.

Num.	Horse Serial Number	Isolate	Origin	Bacterial ID	AMC	IMP	ENR	CHL	GEN	AMK	TMS	MDR
1	1	1.1.1		*Escherichia coli*	2	0	1	2	2	0	0	1
2	2	2.1.1		*Escherichia coli*	2	0	0	2	2	1	2	1
3	3	3.1.1		*Escherichia coli*	2	0	0	2	2	1	2	1
4	6	6.1.1		*Citrobacter sedlakii*	0	0	0	0	0	0	0	0
5	7	7.1.1		*Klebsiella pneumoniae*	2	0	1	2	2	0	2	1
6	15	15.1.1		*Escherichia coli*	2	0	1	2	2	1	2	1
7	15	15.1.2		*Klebsiella pneumoniae*	2	0	2	0	2	1	2	1
8	17	17.1.2		*Klebsiella pneumoniae*	2	0	1	0	0	0	2	1
9	22	22.1.1		*Escherichia coli*	2	0	0	2	2	1	2	1
10	22	22.1.2		*Klebsiella pneumoniae*	2	0	1	0	2	0	2	1
11	31	31.1.1		*Escherichia coli*	1	0	0	2	2	1	2	1
12	32	32.1.1		*Escherichia coli*	2	0	1	2	0	0	2	1
13	46	46.1.1		*Enterobacter cloacae*	2	0	0	0	0	0	0	0
14	60	60.1.1	On admission	*Escherichia coli*	2	0	1	0	0	0	2	1
15	74	74.1.2		*Enterobacter cloacae*	2	0	2	2	2	2	2	1
16	77	77.1.1		*Escherichia coli*	2	0	1	0	0	0	2	1
17	81	81.1.1		*Escherichia coli*	2	0	1	2	2	0	2	1
18	101	101.1.1		*Escherichia coli*	2	0	2	2	2	0	2	1
19	101	101.1.2		*Klebsiella pneumoniae*	2	0	1	0	0	0	2	0
20	107	107.1.1		*Escherichia coli*	2	0	1	0	0	0	2	1
21	112	112.1.1		*Enterobacter cloacae*	2	0	0	1	0	0	0	0
22	113	113.1.1		*Escherichia coli*	2	0	1	0	0	0	0	0
23	120	120.1.1		*Escherichia coli*	1	0	2	2	2	1	2	1

Table A1. *Cont.*

Num.	Horse Serial Number	Isolate	Origin	Bacterial ID	AMC	IMP	ENR	CHL	GEN	AMK	TMS	MDR
24	121	121.1.1		*Escherichia coli*	2	1	2	2	0	0	0	1
25	136	136.1.1		*Escherichia coli*	2	0	2	2	0	0	2	1
26	136	136.1.2		*Klebsiella pneumoniae*	2	0	2	0	2	1	2	1
27	144	144.1.1		*Escherichia coli*	2	0	1	2	2	1	2	1
28	153	153.1.1		*Escherichia coli*	2	0	1	0	0	0	2	0
29	162	162.1.1		*Escherichia coli*	2	0	0	0	0	1	0	1
30	176	176.1.1		*Escherichia coli*	2	0	1	2	2	1	2	1
31	177	177.1.1		*Escherichia coli*	1	0	1	2	2	0	2	1
32	179	179.1.1		*Escherichia coli*	1	0	0	2	2	0	2	1
33	203	203.1.1		*Klebsiella pneumoniae*	0	0	0	0	2	0	0	0
34	239	239.1.1		*Enterobacter cancerogenus*	2	0	0	0	0	0	0	0
35	244	244.1.1		*Citrobacter sedlakii*		0	0	0	0	0		0
36	267	267.1.1		*Escherichia coli*	2	0	1	0	0	0	2	1
37	278	278.1.1		*Escherichia coli*	2	0	1	0	0	0	2	1
38	288	288.1.1		*Escherichia coli*	2	0	1	0	0	0	2	1
39	290	290.1.1		*Escherichia coli*	2	0	1	0	0	0	2	1
40	1	1.2.2		*Klebsiella pneumoniae*	0	0	1	0	2	0	2	1
41	5	5.2.1		*Escherichia coli*	0	0	0	2	2	0	2	1
42	6	6.2.1		*Klebsiella pneumoniae*	1	0	2	0	0	0	2	1
43	6	6.2.2		*Escherichia coli*	0	0	0	0	0	0	2	0
44	7	7.2.1		*Escherichia coli*	1	0	0	0	2	0	2	1
45	7	7.2.2		*Klebsiella pneumoniae*	1	0	2	2	2	0	2	1

Table A1. *Cont.*

Num.	Horse Serial Number	Isolate	Origin	Bacterial ID	AMC	IMP	ENR	CHL	GEN	AMK	TMS	MDR
46	8	8.2.1		*Escherichia coli*	2	0	1	2			2	1
47	8	8.2.2		*Enterobacter cloacae*	2	0	2	2	2	0	2	1
48	15	15.2.1		*Escherichia coli*	1	0	2	2	2	0	2	1
49	15	15.2.2		*Enterobacter cloacae*	2	0	1	2	2	0	2	1
50	16	16.2.1		*Escherichia coli*	0	0	1	2	0	0	2	1
51	16	16.2.2		*Enterobacter cloacae*	2	0	2	2	2	0	2	1
52	29	29.2.1		*Escherichia coli*	1	0	2	2	2	0	2	1
53	29	29.2.2		*Escherichia vulneris*	0	0	1	2	2	0	2	1
54	31	31.2.1		*Escherichia coli*	1	0	2	2	2	0	2	1
55	35	35.2.1		*Klebsiella pneumoniae*	1	0	1	2	2	0	2	1
56	46	46.2.1		*Pantoea spp*	1	0	2	2	2	0		1
57	46	46.2.2		*Escherichia coli*	1	0	2	2	2	0	2	1
58	47	47.2.1		*Escherichia coli*	1	0	2	2	2	0	2	1
59	47	47.2.2		*Enterobacter cloacae*	2	0	2	2	2	2	2	1
60	49	49.2.2		*Enterobacter cloacae*	2	0	1	2	2	2	2	1
61	55	55.2.1		*Escherichia coli*	1		1	2	2	0	2	1
62	55	55.2.2		*Klebsiella pneumoniae*	1	0	2	0	2	0	2	1
63	56	56.2.2		*Enterobacter cloacae*	2	0	2	2	2	1	2	1
64	57	57.2.1		*Escherichia coli*	0	0			0	0	2	0
65	60	60.2.1		*Enterobacter cloacae*	2	0	2	2	2	0	2	1
66	60	60.2.3		*Escherichia coli*	1	0	1	2	2	0	2	1
67	72	72.2.3		*Salmonella group*	1	0	0		2	2	2	1
68	75	75.2.3		*Enterobacter cloacae*	2	0	1	2	2	0	2	1
69	84	84.2.1		*Escherichia coli*	0	0	0	2	2	0	2	1

Table A1. *Cont.*

Num.	Horse Serial Number	Isolate	Origin	Bacterial ID	AMC	IMP	ENR	CHL	GEN	AMK	TMS	MDR
70	85	85.2.1		*Escherichia coli*	0	0	1	2	2	0	2	1
71	85	85.2.2		*Enterobacter cloacae*	2	0	2	2	2	0	2	1
72	87	87.2.1		*Escherichia coli*	0	0	2	2	0	0	2	1
73	87	87.2.2		*Escherichia coli*	1	0	2	2	2	0	2	1
74	89	89.2.1		*Escherichia coli*	1	0	2	2	2	0	2	1
75	89	89.2.2		*Klebsiella pneumoniae*	1	0	2	0	2	0	2	1
76	91	91.2.1		*Escherichia coli*	1	0	2	2	2	2	2	1
77	91	91.2.2		*Enterobacter cloacae*	2	0	2	2	2	0	2	1
78	101	101.2.1		*Escherichia coli*	1	0	2	2	1	0	2	1
79	101	101.2.2		*Klebsiella pneumoniae*	0	0	1	2	0	0	2	1
80	107	107.2.1		*Escherichia coli*	1	0	2	2	2	0	2	1
81	107	107.2.2		*Enterobacter cloacae*	2	0	2	2	2	2	2	1
82	107	107.2.4		*Enterobacter cloacae*	2	0	1	2	2	0	2	1
83	108	108.2.2		*Enterobacter cloacae*	2	0	1	2	2	0	2	1
84	113	113.2.1		*Escherichia coli*			2	2				
85	115	115.2.1		*Escherichia coli*	0	0	2	2	2	0	2	1
86	115	115.2.2		*Citrobacter freundii*	2	0	2	2	2	0	2	1
87	124	124.2.1		*Escherichia coli*	0	0	2	2	2	0	2	1
88	124	124.2.3		*Salmonella enterica*	1	0			2	2	2	1
89	126	126.2.2		*Citrobacter brakii*	2	0	2	2	2	1	2	1
90	127	127.2.1		*Escherichia coli*	1	0	2	2	2	1	2	1
91	127	127.2.2		*Enterobacter cloacae*	2	0	1	2	2	0	2	1
92	136	136.2.1		*Escherichia coli*	1	0	2	2	2	1	2	1

Table A1. *Cont.*

Num.	Horse Serial Number	Isolate	Origin	Bacterial ID	AMC	IMP	ENR	CHL	GEN	AMK	TMS	MDR
93	136	136.2.2		*Klebsiella pneumoniae*	1	0	2	0	2	0	2	1
94	143	143.2.1		*Escherichia coli*	0	0	0	2	2	0	2	1
95	143	143.2.2		*Citrobacter freundii*	2	0	0	2	2	0	2	1
96	144	144.2.1		*Escherichia coli*	2	0	1	2	2	0	2	1
97	144	144.2.2		*Citrobacter freundii*	2	0			2	1	2	1
98	144	144.2.3		*Proteus mirabilis*	1	0			2	0	0	1
99	148	148.2.1		*Escherichia coli*	2	0	0	0	0	0	2	1
100	149	149.2.1		*Escherichia coli*	0	0	0	2	2	0	2	1
101	149	149.2.2		*Enterobacter cloacae*	2	0	1	2	2	2	2	1
102	152	152.2.2		*Escherichia coli*	1	0	2	0	2	0	2	1
103	156	156.2.1		*Enterobacter cloacae*	2	0	1	2	2	2	2	1
104	156	156.2.2		*Escherichia coli*	1	0	2	2	2	0	2	1
105	158	158.2.2		*Klebsiella pneumoniae*	1	0	2	0	2	0	2	1
106	161	161.2.1		*Escherichia coli*	1	0	2	2	2	0	2	1
107	161	161.2.2		*Enterobacter cloacae*	2	0	1	2	2	0	2	1
108	167	167.2.1		*Klebsiella pneumoniae*	1	0	1	2	2	0	2	1
109	176	176.2.1		*Escherichia coli*	1	0	0	2	2	0	2	1
110	177	177.2.1		*Escherichia coli*	1	0	1	2	2	0	2	1
111	177	177.2.2		*Enterobacter cloacae*	2	0	1	2	2	0	2	1
112	181	181.2.1		*Enterobacter cloacae*	2	0	1			2	1	2	1
113	181	181.2.2		*Escherichia coli*	1	0	2	2	2	0	2	1
114	183	183.2.1		*Klebsiella pneumoniae*	1	0	1	2	2	0	2	1
115	183	183.2.2		*Escherichia coli*	1	0	0	2	2	0	2	1

Table A1. *Cont.*

Num.	Horse Serial Number	Isolate	Origin	Bacterial ID	AMC	IMP	ENR	CHL	GEN	AMK	TMS	MDR
116	195	195.2.1		*Escherichia coli*	1	0	1	2	2	0	2	1
117	212	212.2.1		*Escherichia coli*	0	0	1		2	0	0	0
118	216	216.2.1		*Escherichia coli*	0	0	2	2	0	0	2	0
119	219	219.2.1		*Escherichia coli*	1	0	2		2	0	2	1
120	222	222.2.1		*Klebsiella pneumoniae*	1	0	1	2	2	0	2	1
121	222	222.2.2		*Escherichia coli*	0	0	2	2	0	0	2	1
122	223	223.2.1		*Escherichia coli*	1	0	2	2	2	0	0	1
123	224	224.2.1		*Escherichia coli*	1	0	2	2	2	0	2	1
124	224	224.2.2		*Klebsiella pneumoniae*	1	0	2	2	2	0	2	1
125	228	228.2.1		*Klebsiella pneumoniae*	0	0	2	0	2	0	0	1
126	229	229.2.1		*Escherichia coli*	1	0	2	2	2	2	2	1
127	229	229.2.2		*Salmonella enterica*	1	0	2	2	2	2	2	1
128	234	234.2.1		*Raoultella ornithinolytica*	2	0	1	2	2	2	2	1
129	237	237.2.1		*Escherichia coli*	0	0	0	2	0	0	2	1
130	238	238.2.1		*Klebsiella pneumoniae*	1	0	1	0	2	0	2	1
131	243	243.2.1		*Escherichia coli*	0	0	2	2	0	0	2	1
132	243	243.2.2		*Enterobacter cloacae*	2	0	1	2	2	0	2	1
133	246	246.2.1		*Escherichia coli*	1	0			1	0	2	1
134	246	246.2.2		*Enterobacter cloacae*	2	0	2	2	2	0	2	1
135	265	265.2.1		*Enterobacter cloacae*	2	0	1	2	2	0	2	1
136	272	272.2.1		*Enterobacter cloacae*	2	0	1	2	2	0	2	1
137	273	273.2.1		*Enterobacter cloacae*	2	0	1	2	2	0	2	1
138	278	278.2.1		*Escherichia coli*	0	0	1	2	0	0	2	1

Table A1. *Cont.*

Num.	Horse Serial Number	Isolate	Origin	Bacterial ID	AMC	IMP	ENR	CHL	GEN	AMK	TMS	MDR
139	278	278.2.2		*Citrobacter sedlakii*	0	0	2	2	0	0		1
140	278	278.2.4		*Klebsiella pneumoniae*	0	0	2	2	2	0	0	1
141	279	279.2.1		*Klebsiella oxytoca*	2	0	0	0	2	1	2	1
142	279	279.2.2		*Escherichia coli*	0	0	1	2	0	0	2	1
143	289	289.2.1		*Escherichia coli*	1	0	2	0	2	0	2	1
144	H40	H40.2		*Escherichia coli*	2	0	1	0	0	0	2	1
145	H42	H42.1		*Escherichia coli*	2	0	1	2	2	1	2	1
146	H44	H44.1		*Escherichia coli*	2	0	1	2	2	1	2	1
147	H45	H45.2		*Citrobacter farmeri*	1	0	1	2	2	0	2	1
148	H48	H48.2		*Escherichia coli*	2	0	2	2	2	0	2	1
149	H48	H48.3		*Enterobacter cloacae*	2	0	1	2	2	1	2	1
150	H53	H53.1		*Escherichia coli*	2	0	1	0	0	0	2	1
151	H53	H53.2		*Enterobacter cloacae*	2	0	1	2	2	1	2	1
152	H54	H54.1	Farms	*Escherichia coli*	2	0	2	2	2	0	2	1
153	H56	H56.1		*Escherichia coli*	2	0	1	2	2	1	2	1
154	H56	H56.2		*Enterobacter cloacae*	2	0	1	2	2	1	2	1
155	H57	H57.1		*Enterobacter cloacae*	2	0	1	2	2	1	2	1
156	H57	H57.2		*Escherichia coli*	2	0	1	2	2	1	2	1
157	H60	H60.2		*Escherichia coli*	2	0	1	0	0	0	2	1
158	H110	H110.1		*Enterobacter cloacae*	2	0	0	0	0	0	0	0
159	H138	H138.1		*Escherichia coli*	2	0	0	0	0	0	0	1
160	H140	H140.1		*Escherichia coli*	2	0	0	2	2	1	2	1
161	H154	H154.2		*Klebsiella pneumoniae*	2	0	1	1	2	0	0	1

Table A1. *Cont.*

Num.	Horse Serial Number	Isolate	Origin	Bacterial ID	AMC	IMP	ENR	CHL	GEN	AMK	TMS	MDR
162	H157	H157.2		*Klebsiella pneumoniae*	2	0	2	0	0	0	2	1
163	H230	H230.1		*Escherichia coli*	1	0	1	2	2	0	2	1
164	H230	H230.2		*Enterobacter cloacae*	2	0	1	2	2	0	0	1
165	H231	H231.1		*Escherichia coli*	0	0	1	0	0	0	0	0
166	H233	H233.1		*Escherichia coli*	1	0	1	2	2	0	2	1
168	H234	H234.1		*Escherichia coli*	0	0	1	2	2	0	2	1
169	H234	H234.2		*Enterobacter cloacae*	2	0	1	2	2	0	2	1
170	H235	H235.1		*Escherichia coli*	1	0	1	2	2	0	2	1
171	H236	H236.1		*Escherichia coli*	0	0	1	2	2	0	2	1
172	H237	H237.1		*Escherichia coli*	1	0	1	2	2	0	2	1
173	H238	H238.1		*Escherichia coli*	0	0	1	2	2	0	2	1
174	H241	H241.1		*Escherichia coli*	1	0	1	2	2	0	2	1
175	H242	H242.1		*Escherichia coli*	1	0	1	2	2	0	2	1
176	H243	H243.1		*Escherichia coli*	1	0	1	2	2	0	2	1
177	H245	H245.1		*Escherichia coli*	1	0	1	2	2	0	2	1
178	H246	H246.1		*Escherichia coli*	1	0	1	2	2	0	2	1
179	H247	H247.1		*Escherichia coli*	0	0	1	2	2	0	2	1
180	H248	H248.1		*Escherichia coli*	0	0	1	2	2	0	2	1
181	H250	H250.1		*Escherichia coli*	0	0	1	2	2	0	2	1
182	H251	H251.1		*Escherichia coli*	0	0	1	2	2	0	2	1
183	H253	H253.1		*Escherichia coli*	0	0	1	2	2	0	2	1
184	H254	H254.1		*Escherichia coli*	0	0	1	2	2	0	2	1
185	H256	H256.1		*Escherichia coli*	0	0	1	0	0	0	2	1

Table A1. *Cont.*

Num.	Horse Serial Number	Isolate	Origin	Bacterial ID	AMC	IMP	ENR	CHL	GEN	AMK	TMS	MDR
186	H257	H257.1		*Escherichia coli*	1	0	0	1	2	0	2	1
187	H258	H258.1		*Escherichia coli*	0	0	0	0	0	0	2	0
188	H259	H259.1		*Escherichia coli*	0	0	0	0	0	0	2	0
189	H263	H263.1		*Escherichia coli*	0	0	0	0	0	0	2	0
190	H265	H265.1		*Escherichia coli*	1	0	0	1	2	0	2	1
191	H267	H267.1		*Escherichia coli*	0	0	1	0	0	0	2	1
192	H268	H268.1		*Escherichia coli*	1	0	0	1	2	0	2	1

Susceptible = 0, intermediate susceptibility = 1, resistant = 2. Empty cells mean lack of susceptibility test results due to technical reasons.

Table 2. Results of univariable analysis of variables gleaned from the medical records (horses on admission and during hospitalization) and owners' questionnaires (farm horses). Variables were evaluated for association with the outcome of ESBL-E shedding status of the individual animal.

Population Studied	Variable	Classification	*p*-Value
Farm horses	Breed	Quarter Horse Arabian Pacer Warmblood Pony Local	<0.0001
	Sex	Female Male Gelding	0.027
	Farm	Numbered 1–13	
	Hospitalization within the previous year	Yes/No	0.018
	Antibiotic treatment within the previous year	Yes/No	<0.0001
	Age	Ranged from 0.1–23 y	<0.0001
	Time in farm	Ranged from 0–23 y	0.36
On admission	Breed	Quarter Horse Arabian Tennessee Walking horse Friesian Mangalarga Marchador Warmblood Thoroughbred Miniature horse Haflinger Hannoverian Single footed horse Missouri Fox Trotter	0.394
	Age	Years	0.259
	Sex	Female Male Gelding	0.117
	Geographical origin (within the country)	North South Center	0.879
	Hospitalization within the previous year	Yes/No	0.295
	Clinical signs on admission	Gastro-intestinal disorder Neonatology disorder Ophthalmic disorder Reproduction Orthopedic disorder Hematological disorder Respiratory disorder Endocrine disorder Healthy (mares of sick hospitalized foals)	0.587

Table 2. *Cont.*

Population Studied	Variable	Classification	*p*-Value
	Length of illness before admission	Days	0.618
	Antibiotic treatment within the previous year	Yes/No	0.587
	Length of stay	Days	0.169
	Admission charge	-	0.056
During hospitalization	Shedding on admission	Yes/No	0.9
	Clinical signs on admission	Gastro-intestinal disorder Neonatology disorder Ophthalmic disorder Reproduction Orthopedic disorder Hematological disorder Respiratory disorder Endocrine disorder Tumor Teeth lesion Healthy (mares of sick hospitalized foals)	0.428
	Antibiotic treatment during hospitalization	Yes/No	0.841
	Outcome	Discharged/Died	0.174
	Length of stay	Days	0.29
	Admission charge	-	0.69

Table 3. Risk factor analysis for ESBL-E shedding by horses on admission to hospital (logistic regression).

Risk Factor	*p*-Value	OR
Sex (reference: mare)	0.647	
Stallion	0.409	0.571 (95% CI 0.151–2.162)
Gelding	0.639	0.765 (95% CI 0.25–2.34)
Length of stay	0.766	1 (95% CI 0.997–1)
Admission charge	0.184	1 (95% 1–1)

References

1. Vo, A.T.T.; van Duijkeren, E.; Fluit, A.C.; Gaastra, W. Characteristics of extended-spectrum cephalosporin-resistant Escherichia coli and Klebsiella pneumoniae isolates from horses. *Vet. Microbiol.* **2007**, *124*, 248–255. [CrossRef]
2. Paterson, D.L.; Bonomo, R.A. Extended-Spectrum β-Lactamases: A Clinical Update. *Clin. Microbiol. Rev.* **2005**, *18*, 657–686. [CrossRef]
3. Denkel, L.A.; Schwab, F.; Kola, A.; Leistner, R.; Garten, L.; von Weizsäcker, K.; Geffers, C.; Gastmeier, P.; Piening, B. The mother as most important risk factor for colonization of very low birth weight (VLBW) infants with extended-spectrum β-lactamase-producing Enterobacteriaceae (ESBL-E). *J. Antimicrob. Chemother.* **2014**, *69*, 2230–2237. [CrossRef]
4. Schwaber, M.J.; Navon-Venezia, S.; Kaye, K.S.; Ben-Ami, R.; Schwartz, D.; Carmeli, Y. Clinical and economic impact of bacteremia with extended- spectrum-beta-lactamase-producing Enterobacteriaceae. *Antimicrob. Agents Chemother.* **2006**, *50*, 1257–1262. [CrossRef]

5. Ewers, C.; Stamm, I.; Pfeifer, Y.; Wieler, L.H.; Kopp, P.A.; Schønning, K.; Prenger-Berninghoff, E.; Scheufen, S.; Stolle, I.; Günther, S.; et al. Clonal spread of highly successful ST15-CTX-M-15 Klebsiella pneumoniae in companion animals and horses. *J. Antimicrob. Chemother.* **2014**, *69*, 2676–2680. [CrossRef]

6. Ewers, C.; Bethe, A.; Stamm, I.; Grobbel, M.; Kopp, P.A.; Guerra, B.; Stubbe, M.; Doi, Y.; Zong, Z.; Kola, A.; et al. CTX-M-15-D-ST648 Escherichia coli from companion animals and horses: Another pandemic clone combining multiresistance and extraintestinal virulence? *J. Antimicrob. Chemother.* **2014**, *69*, 1224–1230. [CrossRef]

7. Shnaiderman-Torban, A.; Navon-Venezia, S.; Dahan, R.; Dor, Z.; Taulescu, M.; Paitan, Y.; Edery, N.; Steinman, A. CTX-M-15 Producing Escherichia coli Sequence Type 361 and Sequence Type 38 Causing Bacteremia and Umbilical Infection in a Neonate Foal. *J. Equine Vet. Sci.* **2020**, *85*, 102881. [CrossRef]

8. Shnaiderman-Torban, A.; Paitan, Y.; Arielly, H.; Kondratyeva, K.; Tirosh-Levy, S.; Abells-Sutton, G.; Navon-Venezia, S.; Steinman, A. Extended-Spectrum β-Lactamase-Producing Enterobacteriaceae in Hospitalized Neonatal Foals: Prevalence, Risk Factors for Shedding and Association with Infection. *Animals* **2019**, *9*, 600. [CrossRef]

9. Johns, I.; Verheyen, K.; Good, L.; Rycroft, A. Antimicrobial resistance in faecal Escherichia coli isolates from horses treated with antimicrobials: A longitudinal study in hospitalised and non-hospitalised horses. *Vet. Microbiol.* **2012**, *159*, 381–389. [CrossRef]

10. Dolejska, M.; Duskova, E.; Rybarikova, J.; Janoszowska, D.; Roubalova, E.; Dibdakova, K.; Maceckova, G.; Kohoutova, L.; Literak, I.; Smola, J.; et al. Plasmids carrying blaCTX-M-1 and qnr genes in Escherichia coli isolates from an equine clinic and a horseback riding centre. *J. Antimicrob. Chemother.* **2011**, *66*, 757–764. [CrossRef]

11. Maddox, T.W.; Clegg, P.D.; Diggle, P.J.; Wedley, A.L.; Dawson, S.; Pinchbeck, G.L.; Williams, N.J. Cross-sectional study of antimicrobial-resistant bacteria in horses. Part 1: Prevalence of antimicrobial-resistant Escherichia coli and methicillin-resistant Staphylococcus aureus. *Equine Vet. J.* **2012**, *44*, 289–296. [CrossRef]

12. Kaspar, U.; von Lützau, K.; Schlattmann, A.; Rösler, U.; Köck, R.; Becker, K. Zoonotic multidrug-resistant microorganisms among non-hospitalized horses from Germany. *One Health* **2019**, *7*, 100091. [CrossRef]

13. De Lagarde, M.; Larrieu, C.; Praud, K.; Schouler, C.; Doublet, B.; Sallé, G.; Fairbrother, J.M.; Arsenault, J. Prevalence, risk factors, and characterization of multidrug resistant and extended spectrum β-lactamase/AmpC β-lactamase producing Escherichia coli in healthy horses in France in 2015. *J. Vet. Intern. Med.* **2019**, *33*, 902–911. [CrossRef]

14. Maddox, T.W.; Pinchbeck, G.L.; Clegg, P.D.; Wedley, A.L.; Dawson, S.; Williams, N.J. Cross-sectional study of antimicrobial-resistant bacteria in horses. Part 2: Risk factors for faecal carriage of antimicrobial-resistant Escherichia coli in horses. *Equine Vet. J.* **2012**, *44*, 297–303. [CrossRef]

15. Murk, J.-L.A.N.; Heddema, E.R.; Hess, D.L.J.; Bogaards, J.A.; Vandenbroucke-Grauls, C.M.J.E.; Debets-Ossenkopp, Y.J. Enrichment broth improved detection of extended-spectrum-beta-lactamase-producing bacteria in throat and rectal surveillance cultures of samples from patients in intensive care units. *J. Clin. Microbiol.* **2009**, *47*, 1885–1887. [CrossRef]

16. Clinical and Laboratory Standards Institute (CLSI). *Performance Standards for Antimicrobial Susceptibility Testing*, 26th ed.; Clinical and Laboratory Standards Institute: Wayne, PA, USA, 2016.

17. Falagas, M.E.; Karageorgopoulos, D.E. Pandrug Resistance (PDR), Extensive Drug Resistance (XDR), and Multidrug Resistance (MDR) among Gram-Negative Bacilli: Need for International Harmonization in Terminology. *Clin. Infect. Dis.* **2008**, *46*, 1121–1122. [CrossRef]

18. Woodford, N.; Fagan, E.J.; Ellington, M.J. Multiplex PCR for rapid detection of genes encoding CTX-M extended-spectrum β-lactamases. *J. Antimicrob. Chemother.* **2006**, *57*, 154–155. [CrossRef]

19. Lin, S.-P.; Liu, M.-F.; Lin, C.-F.; Shi, Z.-Y. Phenotypic detection and polymerase chain reaction screening of extended-spectrum β-lactamases produced by Pseudomonas aeruginosa isolates. *J. Microbiol. Immunol. Infect.* **2012**, *45*, 200–207. [CrossRef]

20. Tofteland, S.; Haldorsen, B.; Dahl, K.H.; Simonsen, G.S.; Steinbakk, M.; Walsh, T.R.; Sundsfjord, A. Norwegian ESBL Study Group Effects of phenotype and genotype on methods for detection of extended-spectrum-beta-lactamase-producing clinical isolates of Escherichia coli and Klebsiella pneumoniae in Norway. *J. Clin. Microbiol.* **2007**, *45*, 199–205. [CrossRef]

21. Johnson, J.R.; Clermont, O.; Johnston, B.; Clabots, C.; Tchesnokova, V.; Sokurenko, E.; Junka, A.F.; Maczynska, B.; Denamur, E. Rapid and Specific Detection, Molecular Epidemiology, and Experimental Virulence of the O16 Subgroup within Escherichia coli Sequence Type 131. *J. Clin. Microbiol.* **2014**, *52*, 1358–1365. [CrossRef]

22. Ahmed, M.O.; Clegg, P.D.; Williams, N.J.; Baptiste, K.E.; Bennett, M. Antimicrobial resistance in equine faecal Escherichia coli isolates from North West England. *Ann. Clin. Microbiol. Antimicrob.* **2010**, *9*, 12. [CrossRef]

23. Bilavsky, E.; Temkin, E.; Lerman, Y.; Rabinovich, A.; Salomon, J.; Lawrence, C.; Rossini, A.; Salvia, A.; Samso, J.V.; Fierro, J.; et al. Risk factors for colonization with extended-spectrum beta-lactamase-producing enterobacteriaceae on admission to rehabilitation centres. *Clin. Microbiol. Infect. Off. Publ. Eur. Soc. Clin. Microbiol. Infect. Dis.* **2014**, *20*, O804–O810. [CrossRef]

24. Shitrit, P.; Reisfeld, S.; Paitan, Y.; Gottesman, B.-S.; Katzir, M.; Paul, M.; Chowers, M. Extended-spectrum beta-lactamase-producing Enterobacteriaceae carriage upon hospital admission: Prevalence and risk factors. *J. Hosp. Infect.* **2013**, *85*, 230–232. [CrossRef]

25. Ewers, C.; Grobbel, M.; Stamm, I.; Kopp, P.A.; Diehl, I.; Semmler, T.; Fruth, A.; Beutlich, J.; Guerra, B.; Wieler, L.H.; et al. Emergence of human pandemic O25:H4-ST131 CTX-M-15 extended-spectrum-β-lactamase-producing Escherichia coli among companion animals. *J. Antimicrob. Chemother.* **2010**, *65*, 651–660. [CrossRef]

26. Isgren, C.M.; Edwards, T.; Pinchbeck, G.L.; Winward, E.; Adams, E.R.; Norton, P.; Timofte, D.; Maddox, T.W.; Clegg, P.D.; Williams, N.J. Emergence of carriage of CTX-M-15 in faecal Escherichia coli in horses at an equine hospital in the UK; increasing prevalence over a decade (2008–2017). *BMC Vet. Res.* **2019**, *15*, 268. [CrossRef]

27. Maddox, T.W.; Williams, N.J.; Clegg, P.D.; O'Donnell, A.J.; Dawson, S.; Pinchbeck, G.L. Longitudinal study of antimicrobial-resistant commensal Escherichia coli in the faeces of horses in an equine hospital. *Prev. Vet. Med.* **2011**, *100*, 134–145. [CrossRef]

28. European Medicines Agency. Available online: https://www.ema.europa.eu/en/documents/scientific-guideline/reflection-paper-risk-antimicrobial-resistance-transfer-companion-animals_en.pdf (accessed on 15 January 2015).

29. Dunkel, B.; Johns, I.C. Antimicrobial use in critically ill horses. *J. Vet. Emerg. Crit. Care* **2015**, *25*, 89–100. [CrossRef]

30. Norris, J.M.; Zhuo, A.; Govendir, M.; Rowbotham, S.J.; Labbate, M.; Degeling, C.; Gilbert, G.L.; Dominey-Howes, D.; Ward, M.P. Factors influencing the behaviour and perceptions of Australian veterinarians towards antibiotic use and antimicrobial resistance. *PLoS ONE* **2019**, *14*, e0223534.

31. Van Spijk, J.N.; Schmitt, S.; Schoster, A. Infections caused by multidrug-resistant bacteria in an equine hospital (2012–2015). *Equine Vet. Educ.* **2019**, *31*, 653–658. [CrossRef]

32. Johns, I.C.; Adams, E.-L. Trends in antimicrobial resistance in equine bacterial isolates: 1999–2012. *Vet. Rec.* **2015**, *176*, 334. [CrossRef]

33. Colodner, R.; Rock, W.; Chazan, B.; Keller, N.; Guy, N.; Sakran, W.; Raz, R. Risk factors for the development of extended-spectrum beta-lactamase-producing bacteria in nonhospitalized patients. *Eur. J. Clin. Microbiol. Infect. Dis. Off. Publ. Eur. Soc. Clin. Microbiol.* **2004**, *23*, 163–167. [CrossRef]

34. Vázquez-Martínez, E.R.; García-Gómez, E.; Camacho-Arroyo, I.; González-Pedrajo, B. Sexual dimorphism in bacterial infections. *Biol. Sex. Differ.* **2018**, *9*, 27. [CrossRef]

35. Adler, A.; Sturlesi, N.; Fallach, N.; Zilberman-Barzilai, D.; Hussein, O.; Blum, S.E.; Klement, E.; Schwaber, M.J.; Carmeli, Y. Prevalence, Risk Factors, and Transmission Dynamics of Extended-Spectrum-β-Lactamase-Producing Enterobacteriaceae: A National Survey of Cattle Farms in Israel in 2013. *J. Clin. Microbiol.* **2015**, *53*, 3515–3521. [CrossRef]

36. McGowan, C. Welfare of Aged Horses. *Animals* **2011**, *1*, 366–376. [CrossRef] [PubMed]

37. Monroe, M.; Whitworth, J.D.; Wharton, T.; Turner, J. Effects of an Equine-Assisted Therapy Program for Military Veterans with Self-Reported PTSD. *Soc. Amp Anim.* **2019**, *1*, 1–14. [CrossRef]

38. Borgi, M.; Loliva, D.; Cerino, S.; Chiarotti, F.; Venerosi, A.; Bramini, M.; Nonnis, E.; Marcelli, M.; Vinti, C.; De Santis, C.; et al. Effectiveness of a Standardized Equine-Assisted Therapy Program for Children with Autism Spectrum Disorder. *J. Autism Dev. Disord.* **2016**, *46*, 1–9. [CrossRef] [PubMed]

39. White-Lewis, S.; Johnson, R.; Ye, S.; Russell, C. An equine-assisted therapy intervention to improve pain, range of motion, and quality of life in adults and older adults with arthritis: A randomized controlled trial. *Appl. Nurs. Res.* **2019**, *49*, 5–12. [CrossRef]

40. Linder, D.E.; Mueller, M.K.; Gibbs, D.M.; Siebens, H.C.; Freeman, L.M. The Role of Veterinary Education in Safety Policies for Animal-Assisted Therapy and Activities in Hospitals and Nursing Homes. *J. Vet. Med. Educ.* **2016**, *44*, 229–233. [CrossRef]

41. Royden, A.; Ormandy, E.; Pinchbeck, G.; Pascoe, B.; Hitchings, M.D.; Sheppard, S.K.; Williams, N.J. Prevalence of faecal carriage of extended-spectrum β-lactamase (ESBL)-producing Escherichia coli in veterinary hospital staff and students. *Vet. Rec. Open* **2019**, *6*, e000307. [CrossRef]

Article

The Occurrence and Characterization of Extended-Spectrum-Beta-Lactamase-Producing *Escherichia coli* Isolated from Clinical Diagnostic Specimens of Equine Origin

Leta Elias [1], David C. Gillis [1], Tanya Gurrola-Rodriguez [1], Jeong Ho Jeon [2], Jung Hun Lee [2], Tae Yeong Kim [2], Sang Hee Lee [2], Sarah A. Murray [1], Naomi Ohta [3], Harvey Morgan Scott [1], Jing Wu [4] and Artem S. Rogovskyy [1,*]

[1] Department of Veterinary Pathobiology, College of Veterinary Medicine and Biomedical Sciences, Texas A&M University, College Station, TX 77845, USA; REevener@tamu.edu (L.E.); cgillis@cvm.tamu.edu (D.C.G.); tanya.gurrola@tamu.edu (T.G.-R.); smurray@cvm.tamu.edu (S.A.M.); hmscott@cvm.tamu.edu (H.M.S.)

[2] National Leading Research Laboratory, Department of Biological Sciences, Myongji University, Yongin, Gyeonggido 17058, Korea; najashin@hanmail.net (J.H.J.); topmanlv@hanmail.net (J.H.L.); kty6523@gmail.com (T.Y.K.); sangheelee@mju.ac.kr (S.H.L.)

[3] Faculty of Veterinary Medicine, Okayama University of Science, Imabari 794-8555, Japan; n-ohta@vet.ous.ac.jp

[4] Clinical Veterinary Microbiology Laboratory, College of Veterinary Medicine and Biomedical Sciences, Texas A&M University, College Station, TX 77845, USA; jwu@cvm.tamu.edu

* Correspondence: arogovskyy@tamu.edu

Received: 25 October 2019; Accepted: 16 December 2019; Published: 21 December 2019

Simple Summary: The spread and development of extended spectrum beta-lactamase (ESBL)-mediated antimicrobial resistance is a significant concern in healthcare with impacts to animal and public health alike. While the occurrence of the ESBL phenotype in *Escherichia coli* has been investigated in depth by numerous studies, there is still a lack of information regarding ESBL-producing bacterial isolates from clinical specimens of equine origin. In this study, we investigated the incidence of ESBL-producing *E. coli* in hospitalized horses. Overall, 207 *E. coli* isolates were analyzed for their antimicrobial susceptibility and 13 ESBL-producing *E. coli* isolates were genotypically characterized. Seven out of the 13 *E. coli* isolates were found to harbor the resistance genes $bla_{CTX-M-1}$ or bla_{SHV-1} and a novel beta-lactamase TEM gene variant, $bla_{TEM-233}$ was discovered. Furthermore, despite being phenotypically susceptible to tested carbapenems, 1 out of 13 *E. coli* isolates was PCR-positive for the carbapenemase gene, bla_{IMP-1}. The latter is an alarming finding because the presence of carbapenemase resistance genes in equine pathogens is extremely rare. In conclusion, equines can be reservoirs for ESBL-producing *Enterobacteriaceae*, and further investigation into this species group is necessary to understand their impact in the spread and development of antibiotic resistance genes.

Abstract: *Escherichia coli* isolates were recovered from clinical specimens of equine patients admitted to the Texas A&M Veterinary Medical Teaching Hospital over a five-year period. Ceftiofur resistance was used as a marker for potential extended-spectrum beta-lactamase (ESBL)-activity, and of the 48 ceftiofur-resistant *E. coli* isolates, 27.08% ($n = 13$) were phenotypically ESBL-positive. Conventional PCR analysis followed by the $_{large-scale}bla$ Finder multiplex PCR detected the ESBL genes, CTX-M-1 and SHV, in seven out of the 13 isolates. Moreover, beta-lactamase genes of TEM-1-type, BER-type (AmpC), and OXA-type were also identified. Sequencing of these genes resulted in identification of a novel TEM-1-type gene, called $bla_{TEM-233}$, and a study is currently underway to determine if this gene confers the ESBL phenotype. Furthermore, this report is the first to have found *E. coli* ST1308 in horses. This subtype, which has been reported in other herbivores, harbored the SHV-type ESBL gene. Finally, one out of 13 *E. coli* isolates was PCR-positive for the carbapenemase gene, bla_{IMP-1} despite the lack of phenotypically proven resistance to imipenem. With the identification of novel ESBL gene variant and the demonstrated expansion of *E. coli* sequence types in equine patients, this study underscores the need for more investigation of equines as reservoirs for ESBL-producing pathogens.

Keywords: equine; ESBL; *Escherichia coli*; *Enterobacteriaceae*; antimicrobial resistance; CTX-M-1; SHV

1. Introduction

In 2014, the World Health Organization (WHO) released a review that affirmed antimicrobial resistance as a major global threat, with a predicted impact of 100 trillion dollars in economic losses and 10 million deaths attributable to resistant bacteria by 2050 [1,2]. The production of beta-lactamases, a rapidly evolving class of hydrolytic enzymes that inactivate beta-lactam antibiotics, is a significant mechanism of antimicrobial resistance against penicillins and cephalosporins [3–5]. The wide application of beta-lactam antibiotics has been considered as a driving factor in the development and spread of extended-spectrum beta-lactamase (ESBL)-conferred resistance in Gram-negative bacterial pathogens such as *Escherichia coli*, *Klebsiella* spp., and *Salmonella* [4–7]. The ESBL-encoding genes (e.g., bla_{CTX-M}, bla_{SHV}) allow these pathogens to produce enzymes that hydrolyze the beta-lactam ring of penicillins, first-, second-, and third-generation cephalosporins, and aztreonam, although ESBL-positive pathogens still remain susceptible to carbapenems and cephamycins [8,9]. To date, ESBL-positive bacteria are rapidly emerging in a variety of host species worldwide and pose a serious threat to public health [1,4,10,11].

Analogous with human medicine, ESBL-production is a pronounced concern in the veterinary field [12–15]. Specifically, ESBL-producing *E. coli* have been reported as a cause of severe infections in horses [14,16]. Moreover, nosocomial transmission of ESBL-positive pathogens of the *Enterobacteriaceae* family between horses has also been discussed [13,14]. Of additional importance, the possibility of cross-species transmission of ESBL-positive bacterial strains directly represents a health hazard for humans, especially equine handlers and veterinary staff [15,17,18]. Despite this significant problem, however, information on the occurrence and genetic characterization of ESBL-positive *E. coli* isolated from horses is lacking [3,19,20]. Of the studies that have investigated ESBL-positive *E. coli* in equines, the isolates were predominantly originated from fecal samples. To date, very few studies have thoroughly evaluated the occurrence of ESBL-positive *E. coli* in diagnostic specimens (other than feces) from equine patients [14,19]. As such, the objective of the present work was to examine the proportion and genetic diversity of ESBL-positive *E. coli* in clinical specimens of equine origin.

2. Materials and Methods

2.1. Sample Collection and Bacterial Identification

A total of 207 *E. coli* isolates were recovered from equine clinical diagnostic specimens submitted to the Texas A&M Veterinary Medical Teaching Hospital (VMTH) between January 1, 2009 and December 31, 2014. Diagnostic specimens fell into the following categories: abscess (e.g., pus, draining tract swabs), abdominal cavity (e.g., peritoneal fluid), blood, colon, ear, liver, spleen, female and male reproductive systems (e.g., cervix, clitoral sinus, uterus, semen, prepuce), lower and upper respiratory tract (e.g., guttural pouch, transtracheal wash, lung), skeletal system (bone/bony sequestrum, hoof, and joint), surgical site (e.g., incision swab, screw), thoracic cavity (e.g., pleural fluid), urinary system (e.g., bladder, urine), and wound.

All samples were processed immediately after they were submitted to the VMTH Clinical Veterinary Microbiology Laboratory (CVML). The isolates were identified as *E. coli* based on Gram stain, colony morphology, and biochemical analyses that included triple sugar agar, lysine iron agar, motility agar, citrate, indole, and urease tests [21]. Where identification was still in question, the RapID™ One System (Remel, Lenexa, KS, USA) was utilized.

2.2. Antimicrobial Susceptibility Testing

E. coli isolates were tested for antimicrobial susceptibility via broth microdilution using commercially available TREK Sensititre™ Systems (Trek Diagnostic Systems, Inc, Oakwood Village, OH, USA). Since the clinical isolates were tested for their susceptibility against various antimicrobial classes as part of veterinary diagnostic service, and that this service was provided over the five-year period, minimum inhibitory concentration (MIC) data were derived from different antimicrobial susceptibility panels and hence the numbers of isolates tested per panel varied. In addition to Sensititre™ COMEQ3F Plate and Sensititre™ Equine EQUIN1F AST Plate (Trek Diagnostic Systems, Thermo Fisher Scientific, Lenexa, KS, USA), some isolates were tested by a newer panel, Sensititre™ NARMS Gram Negative Plate (Trek Diagnostic Systems, Thermo Fisher Scientific, Lenexa, KS, USA). Breakpoints from the most current Clinical and Laboratory Standards Institute (CLSI) guideline M100 were applied to interpret the MIC results [22]. The *E. coli* isolates that were resistant to ceftiofur were also tested using TREK Sensititre™ ESBL Plate (Trek Diagnostic Systems, Thermo Fisher Scientific, Lenexa, KS, USA). The confirmatory testing included both cefotaxime and ceftazidime. *E. coli* isolates were considered ESBL-positive if there was a 3 or greater two-fold concentration decrease in the MIC for cefotaxime or ceftazidime with clavulanic acid as compared to the MIC for the respective antimicrobial agent when tested alone. *E. coli* (ATCC® 25922™) obtained from the American Type Culture Collection (Old Town Manassas, VA, USA) was used as a CLSI control strain.

2.3. Detection and Characterization of Bla Genes

Genomic DNA was isolated from ESBL-positive *E. coli* using QIAprep Spin™ Miniprep kit (Qiagen, Valencia, CA, USA) according to the manufacturer's instruction. Beta-lactamase genes, bla_{SHV}, bla_{TEM}, and bla_{CTX-M} of groups 1, 2, 8, 9, and 10 were then screened by PCR as previously described [23,24]. The primer sequences are provided in Table S1. The amplicons of expected sizes were further sequenced at Molecular Cloning Laboratories (San Francisco, CA, USA) and the sequence results were verified via BLASTn [25].

In addition, a recently developed detection method, $_{large-scale}bla$ Finder ($_{large-scale}bla$ Finder, Dr. ProLab, Inc., Yongin, South Korea) was also utilized [26]. Specifically, a colony of a fresh overnight culture from LB medium plate was inoculated in 20 μL 0.1% Triton X-100 and then heated at 100 °C for 10 min. After centrifugation at 18,000× *g* for 1 min, the supernatant was used as a DNA template for the multiplex PCR. The multiplex mixture containing 1X Solg™ Multiplex PCR Smart mix and 1 Unit of Uracil-DNA glycosylate (SolGent Co., Ltd., Daejeon, South Korea) was mixed with RNase-free water and primer mixture [26]. The final concentration of each primer was 0.2143 μM. Template DNA was then added to the mixture. Amplification was performed under the following thermal cycling conditions: initial denaturation at 95 °C for 5 min; 30 cycles of 95 °C for 30 s, 64 °C for 40 s, and 72 °C for 50 s; and a final elongation step at 72 °C for 7 min. After amplification samples were stored at 4 °C until further analysis. Resultant amplicons were analyzed by electrophoresis on a 2% agarose gel at 100 V for 1 h and with ethidium bromide staining and then sequenced (Molecular Cloning Laboratories, San Francisco, CA, USA).

2.4. Multi-Locus Sequence Typing (MLST)

To determine the sequence type (ST) of each isolate, seven housekeeping genes, *adk*, *fumC*, *gyrB*, *icd*, *mdh*, *purA*, and *recA* were PCR amplified as described [27]. Specifically, amplification was performed under the following conditions: initial denaturation at 95 °C for 2 min; 30 cycles of 95 °C for 2 min, 52 °C for 1 min, and 72 °C for 2 min; and a final elongation step at 72 °C for 5 min. Resulting sequenced amplicons were used to determine bacterial STs by using the *E. coli* MLST Database [28]. All the primer sequences are provided in Table S2.

2.5. E. coli *Phylogroup Identification*

E. coli phylogroup identification of ESBL-positive isolates was performed as described [29] and PCR conditions were as follows: initial denaturation at 95 °C for 5 min; 30 cycles of 95 °C for 30 s, 55 °C for 30 s, and 72 °C for 30 s; and a final elongation step at 72 °C for 7 min. All the primers are provided in Table S3.

2.6. Ethics

In the present study, all the bacterial isolates that were phenotypically and genotypically characterized had been recovered from clinical specimens of equine origin submitted to the VMTH. Since the clinical specimens were analyzed as part of veterinary diagnostic service, no specific approval on the animal subject was required for phenotypical or genotypical characterization of *E. coli* isolates.

3. Results

3.1. Phenotypic Analysis

Antimicrobial susceptibility testing detected resistance to ampicillin in 28.07% of the *E. coli* isolates (48/171; the number of resistant isolates/the total number of isolates tested; Table 1). Among the third-generation cephalosporins, the lowest proportion of resistance was observed for the *E. coli* isolates tested against ceftazidime, with only 3.25% (4/123) of the isolates being resistant to this antimicrobial. In contrast, when tested against ceftriaxone and ceftiofur, the numbers of resistant isolates were significantly higher, 29.17% (14/48) and 27.08% (13/48), respectively. Resistance to cefpodoxime was observed in 13.89% of the isolates tested (5/36). Lastly, resistance to cefoxitin, a second-generation cephalosporin, was only detected in 4.76% of the *E. coli* isolates (4/84; Table 1).

Table 1. Distribution of minimum inhibitory concentrations of clinical *Escherichia coli* isolates of equine origin.

Antimicrobials	Isolates Tested	# Resistant Isolates [a]	% Resistant Isolates	95% CI Lower	95% CI Upper	<0.015	0.015	0.03	0.06	0.125	0.25	0.5	1	2	4	8	16	32	64	128	256	512	1028
Amikacin [b]	123	^													90.24	6.5	0	1.63	1.63				
Amikacin [d]	36	0	0.00	0	9.74 *										86.11	5.56	8.33	0					
Amoxacillin/Clavulanic Acid [c]	48	4	8.33	2.32	19.98									4.17	16.67	31.25	27.08	6.25	2.08				
Amoxacillin/Clavulanic Acid [d]	36	3	8.33	1.75	22.47									0	38.89	52.78	0	5.56	2.78				
Ampicillin [b]	123	27	21.95	14.99	30.31								5.69	33.33	35.77	2.44	0.81	0.81	21.14				
Ampicillin [c]	48	21	43.75	29.48	58.82								8.33	22.92	20.83	4.17	0	0	43.75				
Ampicillin [d]	36	^											5.56	19.44	13.89	0	0	61.11					
Azithromycin [b]	123	^											4.07	15.45	50.41	30.08							
Azithromycin [c]	48	7	14.58	6.07	27.76								4.17	12.5	52.08	14.58	2.08	14.58					
Cefazolin ‡[b]	123	^													91.87	2.44	0	5.69					
Cefazolin ‡[d]	36	^													0	80.56	8.33	11.11					
Cefoxitin [c]	48	4	8.33	2.32	19.98									22.92	47.92	14.58	6.25	0	8.33				
Cefoxitin [d]	36	0	0.00	0	9.74 *									38.89	41.67	11.11	8.33						
Cefpodoxime [d]	36	5	13.89	4.67	29.50									80.56	5.56	2.78	11.11						
Ceftazodime [b]	123	4	3.25	0.89	8.12								95.93	0.81	0	0	1.63	0	0	1.63			
Ceftiofur [b]	123	^									44.72	48.78	0.81	0.81	0.81	4.07							
Ceftiofur [c]	48	13	27.08	15.28	41.85					2.08	18.75	35.42	10.42	4.17	2.08	0	27.08						
Ceftiofur [d]	36	^									50	27.78	5.56	5.56	2.78	8.33							
Ceftriaxone [c]	48	14	29.17	16.95	44.06						62.5	2.08	2.08	4.17	2.08	0	2.08	6.25	2.08	16.67			
Cephalothin [d]	36	^													13.89	27.78	36.11	22.22					
Chloramphenicol [b]	123	23	18.70	12.24	26.72									37.4	37.4	6.5	2.44	16.26					
Chloramphenicol [c]	48	14	29.17	16.95	44.06								6.25	22.92	33.33	8.33	2.08	27.08					
Chloramphenicol [d]	36	^													36.11	38.89	0	25					
Ciprofloxacin [c]	48	14	29.17	16.95	44.06	64.58	2.08	0	0	4.17	0	0	0	29.17									
Doxycycline [b]	123	32	26.02	18.52	34.70								65.04	7.32	1.63	8.13	17.89						
Enrofloxacin [b]	123	^									88.62	0.81	1.63	0	8.94								
Enrofloxacin [d]	36	5	13.89	4.67	29.50							86.11	0	0	0	13.89						
Gentamicin [b]	123	^											69.92	6.5	1.63	0.81	21.14					
Gentamicin [c]	48	18	37.50	23.95	52.65							10.42	39.58	10.42	0	2.08	0	37.5				
Gentamicin [d]	36	^											44.44	8.33	0	0	47.22					
Imipenem [b]	123	0	0.00	0.00	2.95*								100	0	0							
Imipenem [d]	36	0	0.00	0.00	9.74*								100	0	0							
Marbofloxacin [d]	36	^									86.11	2.78	0	0	11.11							
Naladixic Acid [c]	48	14	29.17	16.95	44.06								12.5	39.58	14.58	4.17	0	0	29.17			
Orbifloxacin [d]	36	^										86.11	0	2.78	11.11							
Streptomycin [c]	48	10	20.83	10.47	34.99									4.17	52.08	16.67	6.25	0	20.83			
Sulfisoxazole [c]	48	^															50	8.33	0	2.08	0		39.58	
Tetracycline [b]	123	^											69.92	1.63	0.81	27.64							
Tetracycline [c]	48	24	50.00	35.23	64.77								50	0	0		4.17		45.83				
Tetracycline [d]	36	^											47.22	0	0	52.78							
Ticarcillin [b]	123	^															77.24	0	0	0.81	21.95			
Ticarcillin [d]	36	^															36.11	2.78	5.56	2.78	52.78			
Ticarcillin/Clavulanic acid [b]	123	^															88.62	4.07	3.25	0.81	3.25			
Ticarcillin/Clavulanic acid [d]	36	0	0.00	0.00	9.74*											72.22	16.67	8.33	2.78				
Trimethoprim-Sulfamethoxazole [b]	123	45	36.59	28.09	45.75									62.6	0.81	0	36.59						
Trimethoprim-Sulfamethoxazole [c]	48	19	39.58	25.77	54.73						60.42	0	0	0	0	0	39.58						
Trimethoprim-Sulfamethoxazole [d]	36	^											33.33	5.56	0	61.11							

Resistance profiles of 207 *E. coli* isolates from equine patients of Texas A&M University Teaching Hospital; [a] The interpretation of minimum inhibitory concentration (MIC) was based on the 2019 Clinical and Laboratory Standards Institute (CLSI) guideline M100 (indicated by vertical red bars) unless otherwise specified; [b] Sensititre™ Equine EQUIN1F AST Plate; [c] Sensititre™ NARMS Gram Negative Plate; [d] Sensititre™ COMEQ3F Plate; * one-sided, 97.5% confidence interval; ^ CLSI MIC breakpoint is above the range of the assay; ‡ CLSI breakpoints for oral cefazolin were used to interpret the MIC.

Resistance against chloramphenicol was identified in 21.64% of the equine isolates (37/171). Of note, 100% of the *E. coli* isolates were susceptible to amikacin (0/36), whereas 37.50% (18/48) and 20.83% (10/48) of the isolates were resistant to gentamicin and streptomycin, respectively (Table 1). Resistance to tetracycline was detected in 50.0% of the *E. coli* isolates (24/48), which was approximately twice as high when compared to resistance of the isolates against doxycycline (26.02%; 32/123). Resistance to ciprofloxacin and enrofloxacin was found in 29.17% (14/48) and 13.89% (5/36) of the equine isolates, respectively. Against nalidixic acid, 29.17% of the isolates (14/48) were resistant. Importantly, none of the isolates (0/159) were resistant to imipenem, the carbapenem antimicrobial, which is commonly used as a reserve drug to treat serious infections by ESBL-positive pathogens in humans [30]. When the isolates were tested against trimethoprim-sulfamethoxazole, resistance was observed in 37.43% of the isolates (64/171; Table 1). It is important to note that while the 207 isolates were tested for their antimicrobial susceptibility in total, some drugs that had breakpoints outside of the range of the susceptibility panels were excluded from the analysis despite their respective MIC data are still presented in Table 1.

Resistance to ceftiofur, which was detected in 13 out of 48 *E. coli* isolates tested, was an indicator of potential ESBL activity and prompted further susceptibility testing to confirm the ESBL phenotype (Table 1). All 13 ceftiofur-resistant isolates displayed the ESBL phenotype: three or greater two-fold concentration decrease in an MIC for cefotaxime or ceftazidime in combination with clavulanic acid compared to the MIC of the respective antimicrobial when tested alone [18,26]. Moreover, 46.15% (6/13), 76.92% (10/13), and 84.62% (11/13) of the isolates were resistant to cefepime, ceftriaxone, and cefpodoxime, respectively (Table 2). Testing against cefoxitin showed that two of the 13 isolates were resistant to this cephamycin. This finding suggested that these two isolates also harbored a cephamycinase gene that would confer resistance to cefoxitin while sustaining the ESBL phenotype–susceptibility in the presence of clavulanic acid to the third-generation cephalosporins. Consistently, all of the ESBL-positive isolates exhibited full susceptibility to imipenem and meropenem (Table 2).

Table 2. Antimicrobial susceptibility of Extended-Spectrum Beta-Lactamase (ESBL)-positive *Escherichia coli* isolates of equine origin.

Antimicrobial	MIC (µg/mL)												
	E1A	E2A	E3A	E4A	E4B	E4C	E5A	E6A	E7ARL	E7ADS	E8A	E8B	E9A
Cefazolin [a]	>16	>16	16	>16	16	>16	>16	>16	>16	>16	>16	>16	>16
Cefepime	16	2	≤1	16	≤1	>16	16	8	2	4	16	>16	≤1
Cefotaxime	64	16	0.5	>64	1	>64	>64	>64	8	16	>64	>64	1
Cefotaxime/Clavulanic acid	≤0.12	≤0.12	≤0.12	≤0.12	≤0.12	0.25	≤0.12	≤0.12	≤0.12	≤0.12	≤0.12	8	≤0.12
Cefoxitin	16	≤4	≤4	16	≤4	32	8	8	≤4	≤4	8	>64	8
Cefpodoxime	>32	>32	8	>32	4	>32	>32	>32	>32	>32	>32	>32	4
Ceftazidime	8	4	4	16	16	16	64	0.5	2	2	16	64	16
Ceftazidime/Clavulanic acid	0.25	≤0.12	≤0.12	0.25	≤0.12	0.5	0.5	≤0.12	0.50	0.25	0.25	16	≤0.12
Ceftriaxone	128	32	≤1	128	≤1	>128	>128	32	64	32	128	128	2
Cephalothin	>16	>16	>16	>16	>16	>16	>16	>16	>16	>16	>16	>16	>16
Ciprofloxacin	>2	>2	≤1	>2	>2	>2	>2	>2	≤1	≤1	>2	>2	>2
Imipenem	≤0.5	≤0.5	≤0.5	≤0.5	≤0.5	≤0.5	≤0.5	≤0.5	≤0.5	≤0.5	≤0.5	≤0.5	≤0.5
Meropenem	≤1	≤1	≤1	≤1	≤1	≤1	≤1	≤1	≤1	≤1	≤1	≤1	≤1
Piperacillin-Tazobactam	8	≤4	≤4	16	≤4	16	16	≤4	≤4	≤4	16	16	≤4

[a] The CLSI breakpoint for oral cefazolin was used to interpret the MIC.

3.2. Genetic Characterization of ESBL-Positive E. coli Isolates

In order to detect ESBL and other beta-lactamase genes in the 13 ESBL-positive *E. coli* isolates, in addition to conventional PCR [23,24], a more comprehensive multiplex PCR-based detection method, large-scale*bla* Finder [26] was used in the present study. Moreover, multilocus sequence typing (MLST) analysis was also utilized to determine sequence types of the 13 *E. coli* isolates. As a result, a total of six distinct *bla* gene types were identified: TEM, BER, SHV, CTX-M-1, OXA-1, and IMP (Table 3). Of these, the AmpC beta-lactamase gene, *bla*$_{BER}$, was most represented among the isolates, with its proportion being 84.62% (11/13). A variant of previously characterized *bla*$_{BER}$ gene (GenBank accession number: EF125541) was detected in one of the 13 isolates. This BER-type *bla* variant had five silent mutations, T375C, G378A, C387T, T477G, and T576A. Moreover, additional BER-type mutations were observed in *bla*$_{BER}$ of another *E. coli* isolate. In addition to the five silent mutations of the original BER variant (GenBank accession number: EF125541), this BER variant 1 (BER-v1) had two additional nucleotide substitutions, G469A and G1012A, which translated into their respective amino-acid substitutions, A157T and G338S, in BER (GenBank accession number: ABM69263). Also, the study detected another BER-type variant (BER-v2), whose mutations were also consistent with the 5 silent mutations of the original BER variant (GenBank accession number: EF125541; Table 3). Yet, this BER-v2 had also four additional nucleotide substitutions, G245A, G313A, G469A, and G1012A (GenBank accession number: EF125541), which, respectively, translated into 4 amino-acid substitutions, S28N, A105T, A157T, and G338S (GenBank accession number: ABM69263). The second most commonly identified *bla* gene was TEM-1, which was found in 69.23% (9/13) of the *E. coli* isolates. Two silent mutations, C228T and G396T, in TEM-1 were consistently found in the nine isolates. CTX-M-1 was detected in 30.77% (4/13) of the *E. coli* isolates. Furthermore, SHV-12 and OXA-1 were detected in 23.07% (3/13) of the isolates each. Alarmingly, the metallo-beta-lactamase IMP-1 was detected in one out of the 13 isolates (Table 3). Lastly, a novel TEM-1-type beta-lactamase gene, designated as *bla*$_{TEM-233}$, was detected in isolate E9A52022 (GenBank accession number: MH270416; Table 3).

Table 3. The *bla* genes detected in ESBL-positive *Escherichia coli* isolates of equine origin.

Isolate ID	*bla* Gene Type Detected Using $_{large-scale}$*bla* Finder Kit	*bla* Gene Name by Sequencing of Simplex PCR Products Using Long-Length Primer Pairs of $_{large-scale}$*bla* Finder Kit to Detect Each ORF (GenBank Accession No. of Gene)	Phylogroup	MLST
E1A17025	TEM type	Two silent mutations (C228T and G396T) in *bla*$_{TEM-1}$ (J01749)	D	648
E2A28099DS	TEM type	Two silent mutations (C228T and G396T) in *bla*$_{TEM-1}$ (J01749)	B1	167
	BER type	*bla*$_{BER}$ [a] (EF125541)		
	BER type	*bla*$_{BER}$ [a] (EF125541)		
E3A31074	TEM type	Two silent mutations (C228T and G396T) in *bla*$_{TEM-1}$ (J01749)	B1	1308
	SHV type	*bla*$_{SHV-12}$ [a] (AY008838)		
	BER type	Five silent mutations (T375C, G378A, C387T, T477G, and T576A) in *bla*$_{BER}$ (EF125541)		
E4A39024	BER type	*bla*$_{BER}$ [a] (EF125541)	B2	648
E4B39025	TEM type	Two silent mutations (C228T and G396T) in *bla*$_{TEM-1}$ (J01749)	B1	224
	BER type	*bla*$_{BER}$ [a] (EF125541)		
E4C44009	CTX-M-1 type	*bla*$_{CTX-M-3}$ [a] (AB976577)	D	648
	BER type	*bla*$_{BER}$ [a] (EF125541)		
E5A41032	TEM type	Two silent mutations (C228T and G396T) in *bla*$_{TEM-1}$ (J01749)	B2	410
	SHV type	*bla*$_{SHV-12}$ [a] (AY008838)		
	BER type	*bla*$_{BER-v2}$ (*bla*$_{BER}$ variant 2) with five silent mutations (T375C, G378A, C387T, T477G, and T576A) and four nucleotide substitutions (G245A, G313A, G469A, and G1012A) in *bla*$_{BER}$ (EF125541), which caused four amino acid substitutions (S82N, A105T, A157T, and G338S) in BER (ABM69263) and was called as BER-v2 (BER variant 2)		
	OXA-1 type	*bla*$_{OXA-1}$ [a] (GU119958)		
E6A43048	TEM type	Two silent mutations (C228T and G396T) in *bla*$_{TEM-1}$ (J01749)	D	648
	BER type	*bla*$_{BER-v1}$ (*bla*$_{BER}$ variant 1) with five silent mutations (T375C, G378A, C387T, T477G, and T576A) and two nucleotide substitutions (G469A and G1012A) in *bla*$_{BER}$ (EF125541), which caused two amino acid substitutions (A157T and G338S) in BER (ABM69263) and was called as BER-v1 (BER variant 1)		
E7A44050DS	CTX-M-1 type	*bla*$_{CTX-M-3}$ [a] (AB976577)	A	10
	IMP type	*bla*$_{IMP-1}$ [a] (AB472901)		
	BER type	*bla*$_{BER}$ [a] (EF125541)		
E7A44050RL	BER type	*bla*$_{BER}$ [a] (EF125541)	A	10
E8A49072	TEM type	Two silent mutations (C228T and G396T) in *bla*$_{TEM-1}$ (J01749)	A	410
	CTX-M-1 type	*bla*$_{CTX-M-3}$ [a] (AB976577)		
	BER type	*bla*$_{BER}$ [a] (EF125541)		
	OXA-1 type	*bla*$_{OXA-1}$ [a] (GU119958)		

Table 3. *Cont.*

Isolate ID	*bla* Gene Type Detected Using large-scale *bla* Finder Kit	*bla* Gene Name by Sequencing of Simplex PCR Products Using Long-Length Primer Pairs of large-scale *bla* Finder Kit to Detect Each ORF (GenBank Accession No. of Gene)	Phylogroup	MLST
E8B49043	TEM type	Two silent mutations (C228T and G396T) in bla_{TEM-1} (J01749)	A	410
	CTX-M-1 type	$bla_{CTX-M-3}$ [a] (AB976577)		
	BER type	bla_{BER} [a] (EF125541)		
	OXA-1 type	bla_{OXA-1} [a] (GU119958)		
E9A52022	TEM type	$bla_{TEM-233}$ [a] (MH270416)	B1	156
	SHV type	bla_{SHV-12} [a] (AY008838)		

[a] 100% nucleotide sequence identity to each gene described as GenBank accession number.

3.3. Phylogenetic Grouping

The 13 ESBL-positive *E. coli* isolates belonged to four phylogenetic groups that were represented by a total of seven distinct sequence types (Table 3). Four isolates of ST10 (*n* = 2) and ST410 (*n* = 2) belonged to phylogroup A. Phylogroup B1 was most diverse and included four equine isolates with distinct sequence types: ST167, ST1308, ST224, and ST156. Phylogroup B2 had only two clinical isolates, ST168 and ST410. Lastly, phylogroup D was uniformly represented by three ST648 isolates (Table 3).

4. Discussion

In the present study, a total of 207 *E. coli* isolates were cultured from clinical diagnostic specimens collected from equine patients, which were admitted to the Texas A&M Veterinary Medical Teaching Hospital from 2009 through 2014. The results of the present study demonstrated that 27.08% of the isolates screened with ceftiofur expressed the ESBL-positive phenotype. Recent studies have reported the occurrence of ESBL-producing *E. coli* isolates recovered from equines to range from as low as 0.2% (one out of 508 isolates tested) in feral horses living on an isolated Canadian island [31], to 84% in equine patients at a veterinary teaching hospital in the Netherlands [19]. Additional studies detected ESBL-producing *E. coli* in 6.3% of fecal samples from equine patients across various veterinary practices in the United Kingdom [12], 10.1% of equine patients at a veterinary clinic in Germany [20], and 32% of equine patients at a veterinary clinic in the Czech Republic [3].

Resistance to chloramphenicol was present in a high proportion of the *E. coli* isolates (21.64%; 37/171). Due to its negative side effects, chloramphenicol is banned in human medicine and is considered a last choice drug to treat gastrointestinal disease (e.g., abdominal abscesses and salmonellosis) in horses [32]. Thus, the usage of chloramphenicol in equine medicine may explain the high proportion of resistance detected in the tested isolates.

Overall, the 13 ESBL-positive *E. coli* isolates represented a total of seven sequence types: ST648, ST410, ST10, ST224, ST167, ST1308, and ST156. Of these, ST648, ST167, ST410, ST224, and ST10 have been described as extended host spectrum genotypes [13]. ST648 was the most prevalent sequence type isolated in this study and encompassed a total of four ESBL-positive *E. coli* isolates that were recovered from three equine patients. *E. coli* ST648 is associated multi-drug resistance and high virulence, drawing comparisons with ST131, which is recognized as an internationally relevant high-risk *E. coli* [13,33]. *E. coli* ST648 was recovered from a variety of animals including canines, felines, horses, livestock, wild birds, and humans [34–36]. ST410, which was identified in 3 of the 13 *E. coli* isolates, has also been described as an emerging high-risk *E. coli* with potential international implications [37]. ST410 was previously isolated from humans [38,39], canines, felines [40], swine, poultry, cattle [41–43], as well as birds [44]. *E. coli* ST10, identified in two of the 13 isolates, was recovered from humans, turkey meat, chickens, swine, cattle [45,46], and horses [19,20]. Lastly, ST224, ST167, ST1308, and ST156 were detected in one isolate each. Notably, both ST10 and ST224 previously demonstrated their capacity for nosocomial infections and the spread of these sequence types between horses and potentially, to their human handlers [14,18,19]. In addition to being recovered from horses [14], ST224 was isolated from humans [11], swine [42], bovines [47], birds [44], and in this study, from a donkey. ST167 *E.*

coli was isolated from humans, cattle, swine, wild birds [34,46,48], turkey meat [43], and horses [49]. Furthermore, this sequence type has been associated with the global carriage of ESBL-producing *E. coli* [13,34,48]. *E. coli* ST156 was identified in fish [50], canines, felines, horses [51], chickens, and other avian species [52,53]. Lastly, this study is the first demonstration of ST1380 *E. coli* being isolated from equines, the sequence type that was previously isolated from swine and bovine species [54,55].

Phylogenetic analysis showed that the 13 ESBL-positive *E. coli* isolates belonged to four phylogroups: A, B1, B2, and D. The most represented phylogroups were A and B1, with each encompassing 4 of the 13 ESBL-positive *E. coli* isolates. The phylogroup A, which commonly includes commensal strains of *E. coli* [29], was composed of the ST410 and ST10 isolates. The sequence types, ST224, ST167, ST1308, and ST156 belonged to phylogroup B1. Furthermore, three of the 4 ESBL-positive *E. coli* ST648 isolates fell into phylogroup group D. One of the ESBL-positive *E. coli* ST648 isolates belonged to phylogroup B2, which, in addition to phylogroup D, is most often associated with virulent extraintestinal infections [56]. Finally, phylogroup B2 included one ESBL-positive *E. coli* isolate with the sequence type ST410 [57]. It should be emphasized that despite the fact that phylogroups A and B1 are associated with commensal *E. coli*, which are considered harmless, these organisms can act as reservoirs for ESBL gene-carrying plasmids and, therefore, may contribute to the spread of resistance among pathogenic bacteria [57–59]. This, in turn, puts both humans and animals at risk for the nosocomial spread and cross-species transfer of ESBL resistance genes [13,15,17,18].

The 13 ESBL-positive *E. coli* isolates were screened for bla_{SHV}, bla_{TEM}, and bla_{CTX} of groups 1, 2, 8, 9, and 10. In addition to the conventional PCR-based approach [23,24], the recently developed $_{large-scale}bla$ Finder detection method was utilized to more thoroughly examine the isolates for the presence of most clinically relevant beta-lactamase genes [26]. As a result, a novel TEM-1-type beta-lactamase gene, designated as $bla_{TEM-233}$, was detected in one *E. coli* isolate. Further investigation is needed to determine whether or not this newly identified variant is functional. Moreover, the TEM-1-type *bla* gene with two silent mutations at C228T and G396T was consistently detected in eight out of the 13 ESBL-positive isolates of the following sequence types: ST648, ST167, ST1308, ST224, and ST410. Of these TEM-1 harboring isolates, the most represented phylotype was B1, which included three of the 8 isolates. One *E. coli* isolate represented group B2 and the other four isolates belonged to groups A and D (two in each group). Furthermore, a variant of the AmpC beta-lactamase producing gene [60] was detected in one of the 13 isolates. This BER-type *bla* variant fell within phylogroup B1 and belonged to *E. coli* of ST1308. Additionally, BER-v1 (BER variant 1) was observed in one isolate of phylogroup D and ST648. Lastly, BER-v2 was of ST410 and belonged to phylogroup B2.

It should be noted that despite all 13 isolates that exhibited the ESBL phenotype, only seven isolates were genotypically confirmed to harbor ESBL-resistance genes. While the $_{large-scale}bla$ Finder method identifies a much wider array of clinically relevant *bla* genes compared to the conventional PCR approach, it does not detect all the existing ESBL genes. As such, it is well possible that some ESBL genes remained undetected in the other six isolates. Additionally, the carbapenemase gene, bla_{IMP-1}, was found in isolate E7A44050DS despite the lack of detectable carbapenem resistance when tested phenotypically. The latter was a surprising finding because this carbapenemase gene had no mutations, which was determined through three independent sequencing runs. Previously, it was shown that MICs of carbapenem-producing *Enterobacteriaceae* may vary greatly and even be below the CLSI-established carbapenem breakpoints [61]. Moreover, two out of the seven genotypically-confirmed ESBL-positive isolates were also resistant to cefoxitin, a second-generation cephalosporin, which is not typical of ESBL-producing *E. coli* and is more commonly associated with AmpC beta-lactamase-producing bacteria [8,26,62]. Together, these interesting results warrant further genetic testing (e.g., via whole genome sequencing) for a more thorough analysis of these *E. coli* isolates.

5. Conclusions

In summary, this study demonstrated the first occurrence of *E. coli* ST1380 recovered from clinical specimens of equine origin, a finding that indicates a wider host-range for this *E. coli* ST than was previously reported. This ST1308 *E. coli* isolate harbored the *bla*$_{SHV-12}$ ESBL gene, highlighting the necessity of studying the spread and development of ESBL genes in equines. Alarmingly, one *E. coli* isolate was PCR-positive for the carbapenemase gene, *bla*$_{IMP-1}$ despite this isolate was phenotypically susceptible to imipenem. Lastly, as a result of genetic characterization of beta-lactamase-positive equine isolates, a novel TEM-1-like gene was identified and a study is currently underway to test if this novel ESBL gene is fully functional.

Supplementary Materials: The following are available online at http://www.mdpi.com/2076-2615/10/1/28/s1: **Table S1**. Primers used for screening various types of beta-lactamase genes; **Table S2**. Primers for housekeeping genes (*adk, fumC, gyrB, icd, mdh, purA*, and *recA*) used in MLST analysis; **Table S3**. Primers used for phylogroup determination.

Author Contributions: Conceptualization: A.S.R.; methodology: L.E., D.C.G., T.G.-R., J.H.J., J.H.L., T.Y.K., S.H.L., S.A.M., N.O., H.M.S., J.W., A.S.R.; formal analysis: L.E., J.H.J., J.H.L., T.Y.K., S.H.L., S.A.M., N.O., H.M.S., J.W., A.S.R.; resources: A.S.R., S.H.L., H.M.S.; writing—original draft preparation: L.E., A.S.R.; writing—review and editing: L.E., S.H.L., S.A.M., N.O., H.M.S., A.S.R.; supervision: A.S.R.; funding acquisition: A.S.R., S.H.L., H.M.S. All authors have read and agreed to the published version of the manuscript.

Funding: The genetic characterization of ESBL-producing *E. coli* isolates was supported by the research grant from the Bio and Medical Technology Development Program of the NRF, the Ministry of Science and ICT (MSIT; grant number NRF-2017M3A9E4078014 provided to S.H.L.).

Acknowledgments: The members of the Texas A&M Veterinary Medical Teaching Hospital—Clinical Veterinary Microbiology Laboratory performed initial isolation and antimicrobial susceptibility testing on a portion of the bacterial isolates as part of routine veterinary diagnostic service. The experimental work was supported through Department of Veterinary Pathobiology, Texas A&M College of Veterinary Medicine and Biomedical Sciences start-up package provided to A.S. Rogovskyy.

Conflicts of Interest: The authors declare no conflict of interest. The funders had no role in the design of the study; in the collection, analyses, or interpretation of data; in the writing of the manuscript; or in the decision to publish the results.

References

1.   World Health Organization (WHO). *Antimicrobial Resistance: Global Report on Surveillance;* WHO: Geneva, Switzerland, 2014; pp. 1–256.

2.   O'Neill, J. *Review on Antimicrobial Resistance. Antimicrobial Resistance: Tackling a Crisis for the Health and Wealth of Nations;* Review on Antimicrobial Resistance: London, UK, 2014; pp. 1–20.

3.   Dolejska, M.; Duskova, E.; Rybarikova, J.; Janoszowska, D.; Roubalova, E.; Dibdakova, K.; Maceckova, G.; Kohoutova, L.; Literak, I.; Smola, J.; et al. Plasmids carrying *bla*CTX-M-1 and *qnr* genes in *Escherichia coli* isolates from an equine clinic and a horseback riding centre. *J. Antimicrob. Chemother.* **2011**, *66*, 757–764. [CrossRef]

4.   Rawat, D.; Nair, D. Extended-spectrum beta-lactamases in Gram negative bacteria. *J. Glob. Infect. Dis.* **2010**, *2*, 263–274. [CrossRef] [PubMed]

5.   Livermore, D.M.; Woodford, N. The β-lactamase threat in *Enterobacteriaceae, Pseudomonas* and *Acinetobacter*. *Trends Microbiol.* **2006**, *14*, 413–420. [CrossRef] [PubMed]

6.   Bradford, P.A. Extended-spectrum β-lactamases in the 21st century: Characterization, epidemiology, and detection of this important resistance threat. *Clin. Microbiol. Rev.* **2001**, *14*, 933. [CrossRef] [PubMed]

7.   Rudresh, S.M.; Nagarathnamma, T. Extended spectrum β-lactamase producing *Enterobacteriaceae* & antibiotic co-resistance. *Indian J. Med. Res.* **2011**, *133*, 116–118. [PubMed]

8.   Paterson, D.L.; Bonomo, R.A. Extended-spectrum β-lactamases: A clinical update. *Clin. Microbiol. Rev.* **2005**, *18*, 657–686. [CrossRef]

9.   Datta, N.; Kontomichalou, P. Penicillinase synthesis controlled by infectious *R* factors in *Enterobacteriaceae*. *Nature* **1965**, *208*, 239–241. [CrossRef]

10. Pitout, J.D. Infections with extended-spectrum beta-lactamase-producing *Enterobacteriaceae*: Changing epidemiology and drug treatment choices. *Drugs* **2010**, *70*, 313–333. [CrossRef]

11. Mshana, S.E.; Imirzalioglu, C.; Hain, T.; Domann, E.; Lyamuya, E.F.; Chakraborty, T. Multiple ST clonal complexes, with a predominance of ST131, of *Escherichia coli* harbouring *bla*CTX-M-15 in a tertiary hospital in Tanzania. *Clin. Microbiol. Infect.* **2011**, *17*, 1279–1282. [CrossRef]

12. Maddox, T.W.; Pinchbeck, G.L.; Clegg, P.D.; Wedley, A.L.; Dawson, S.; Williams, N.J. Cross-sectional study of antimicrobial-resistant bacteria in horses. part 2: Risk factors for faecal carriage of antimicrobial-resistant *Escherichia coli* in horses. *Equine Vet. J.* **2012**, *44*, 297–303. [CrossRef]

13. Wieler, L.H.; Ewers, C.; Guenther, S.; Walther, B.; Lubke-Becker, A. Methicillin-resistant staphylococci (MRS) and extended-spectrum beta-lactamases (ESBL)-producing *Enterobacteriaceae* in companion animals: Nosocomial infections as one reason for the rising prevalence of these potential zoonotic pathogens in clinical samples. *Int. J. Med. Microbiol.* **2011**, *301*, 635–641. [PubMed]

14. Walther, B.; Lübke-Becker, A.; Stamm, I.; Gehlen, H.; Barton, A.K.; Janssen, T.; Wieler, L.H.; Guenther, S. Suspected nosocomial infections with multi-drug resistant *E. coli*, including extended-spectrum beta-lactamase (ESBL)-producing strains, in an equine clinic. *Berl. Munch. Tierarztl. Wochenschr.* **2014**, *127*, 421–427. [PubMed]

15. de Lagarde, M.; Larrieu, C.; Praud, K.; Schouler, C.; Doublet, B.; Sallé, G.; Fairbrother, J.M.; Arsenault, J. Prevalence, risk factors, and characterization of multidrug resistant and extended spectrum β-lactamase/AmpC β-lactamase producing *Escherichia coli* in healthy horses in France in 2015. *J. Vet. Intern. Med.* **2019**, *33*, 902–911. [CrossRef] [PubMed]

16. Smet, A.; Boyen, F.; Flahou, B.; Doublet, B.; Praud, K.; Martens, A.; Butaye, P.; Cloeckaert, A.; Haesebrouck, F. Emergence of CTX-M-2-producing *Escherichia coli* in diseased horses: Evidence of genetic exchanges of *bla*CTX-M-2 linked to ISCR1. *J. Antimicrob. Chemother.* **2012**, *67*, 1289–1291. [CrossRef] [PubMed]

17. Chung, Y.S.; Song, J.W.; Kim, D.H.; Shin, S.; Park, Y.K.; Yang, S.J.; Lim, S.K.; Park, K.T.; Park, Y.H. Isolation and characterization of antimicrobial-resistant *Escherichia coli* from national horse racetracks and private horse-riding courses in Korea. *J. Vet. Sci.* **2016**, *17*, 199–206. [CrossRef] [PubMed]

18. Huijbers, P.M.; de Kraker, M.; Graat, E.A.; van Hoek, A.H.; van Santen, M.G.; de Jong, M.C.; van Duijkeren, E.; de Greeff, S.C. Prevalence of extended-spectrum β-lactamase-producing *Enterobacteriaceae* in humans living in municipalities with high and low broiler density. *Clin. Microbiol. Infect.* **2013**, *19*, E256–E259. [CrossRef]

19. Apostolakos, I.; Franz, E.; van Hoek, A.; Florijn, A.; Veenman, C.; Sloet-van Oldruitenborgh-Oosterbaan, M.; Dierikx, C.E. Occurrence and molecular characteristics of ESBL/AmpC-producing *Escherichia coli* in faecal samples from horses in an equine clinic. *J. Antimicrob. Chemother.* **2017**, *72*, 1915–1921. [CrossRef]

20. Walther, B.; Klein, K.S.; Barton, A.K.; Semmler, T.; Huber, C.; Wolf, S.A.; Tedin, K.; Merle, R.; Mitrach, F.; Guenther, S.; et al. Extended-spectrum beta-lactamase (ESBL)-producing *Escherichia coli* and *Acinetobacter baumannii* among horses entering a veterinary teaching hospital: The contemporary "trojan horse". *PLoS ONE* **2018**, *13*, e0191873. [CrossRef]

21. Quinn, P.J.; Markey, B.K.; Leonard, F.C.; Hartigan, P.; Fanning, S.; Fitzpatrick, E. *Veterinary Microbiology and Microbial Disease: Pathogenic Bacteria*; Wiley-Blackwell, Ltd.: Oxford, UK, 2011; pp. 1–286.

22. CLSI. *Performance Standards for Antimicrobial Susceptibility Testing*, 29th ed.; CLSI supplement M100; Clinical and Laboratory Standards Institute: Wayne, PA, USA, 2019.

23. Pitout, J.D.; Thomson, K.S.; Hanson, N.D.; Ehrhardt, A.F.; Moland, E.S.; Sanders, C.C. Beta-lactamases responsible for resistance to expanded-spectrum cephalosporins in *Klebsiella pneumoniae*, *Escherichia coli*, and *Proteus mirabilis* isolates recovered in South Africa. *Antimicrob. Agents Chemother.* **1998**, *42*, 1350–1354. [CrossRef]

24. Woodford, N.; Fagan, E.J.; Ellington, M.J. Multiplex PCR for rapid detection of genes encoding CTX-M extended-spectrum β-lactamases. *J. Antimicrob. Chemother.* **2005**, *57*, 154–155. [CrossRef]

25. Coordinators, N.R. Database resources of the National Center for Biotechnology Information. *Nucleic Acids Res.* **2018**, *46*, D8–D13. [CrossRef] [PubMed]

26. Lee, J.J.; Lee, J.H.; Kwon, D.B.; Jeon, J.H.; Park, K.S.; Lee, C.R.; Lee, S.H. Fast and accurate large-scale detection of beta-lactamase genes conferring antibiotic resistance. *Antimicrob. Agents Chemother.* **2015**, *59*, 5967–5975. [CrossRef] [PubMed]

27. Wirth, T.; Falush, D.; Lan, R.; Colles, F.; Mensa, P.; Wieler, L.H.; Karch, H.; Reeves, P.R.; Maiden, M.C.; Ochman, H.; et al. Sex and virulence in *Escherichia coli*: An evolutionary perspective. *Mol. Microbiol.* **2006**, *60*, 1136–1151. [CrossRef] [PubMed]

28. Alikhan, N.F.; Zhou, Z.; Sergeant, M.J.; Achtman, M. A genomic overview of the population structure of Salmonella. *PLoS Genet.* **2018**, *14*, e1007261. [CrossRef] [PubMed]

29. Clermont, O.; Bonacorsi, S.; Bingen, E. Rapid and simple determination of the *Escherichia coli* phylogenetic group. *Appl. Environ. Microbiol.* **2000**, *66*, 4555–4558. [CrossRef]

30. Osei Sekyere, J.; Govinden, U.; Bester, L.A.; Essack, S.Y. Colistin and tigecycline resistance in carbapenemase-producing Gram-negative bacteria: Emerging resistance mechanisms and detection methods. *J. Appl. Microbiol.* **2016**, *121*, 601–617. [CrossRef]

31. Timonin, M.E.; Poissant, J.; McLoughlin, P.D.; Hedlin, C.E.; Rubin, J.E. A survey of the antimicrobial susceptibility of *Escherichia coli* isolated from Sable Island horses. *Can. J. Microbiol.* **2016**, *63*, 246–251. [CrossRef]

32. Weese, J.; Baptiste, K.; Baverud, V.; Toutain, P.E. *Guidelines for Antimicrobial Use in Horses*, 1st ed.; Blackwell Publishing Ltd.: Ames, IA, USA, 2008.

33. Schaufler, K.; Semmler, T.; Wieler, L.H.; Trott, D.J.; Pitout, J.; Peirano, G.; Bonnedahl, J.; Dolejska, M.; Literak, I.; Fuchs, S.; et al. Genomic and functional analysis of emerging virulent and multidrug-resistant *Escherichia coli* lineage sequence type 648. *Antimicrob. Agents Chemother.* **2019**, *63*, e00243-19. [CrossRef]

34. Ewers, C.; Bethe, A.; Semmler, T.; Guenther, S.; Wieler, L.H. Extended-spectrum beta-lactamase-producing and AmpC-producing *Escherichia coli* from livestock and companion animals, and their putative impact on public health: A global perspective. *Clin. Microbiol. Infect.* **2012**, *18*, 646–655. [CrossRef]

35. Ewers, C.; Bethe, A.; Stamm, I.; Grobbel, M.; Kopp, P.A.; Guerra, B.; Stubbe, M.; Doi, Y.; Zong, Z.; Kola, A.; et al. CTX-M-15-D-ST648 *Escherichia coli* from companion animals and horses: Another pandemic clone combining multiresistance and extraintestinal virulence? *J. Antimicrob. Chemother.* **2014**, *69*, 1224–1230. [CrossRef]

36. Guenther, S.; Grobbel, M.; Beutlich, J.; Bethe, A.; Friedrich, N.D.; Goedecke, A.; Lubke-Becker, A.; Guerra, B.; Wieler, L.H.; Ewers, C. CTX-M-15-type extended-spectrum beta-lactamases-producing *Escherichia coli* from wild birds in Germany. *Environ. Microbiol. Rep.* **2010**, *2*, 641–645. [CrossRef] [PubMed]

37. Roer, L.; Overballe-Petersen, S.; Hansen, F.; Schønning, K.; Wang, M.; Røder, B.L.; Hansen, D.S.; Justesen, U.S.; Andersen, L.P.; Fulgsang-Damgaard, D.; et al. *Escherichia coli* sequence type 410 is causing new international high-risk clones. *mSphere* **2018**, *3*, e00337-18. [CrossRef] [PubMed]

38. Sidjabat, H.E.; Paterson, D.L.; Adams-Haduch, J.M.; Ewan, L.; Pasculle, A.W.; Muto, C.A.; Tian, G.B.; Doi, Y. Molecular epidemiology of CTX-M-producing *Escherichia coli* isolates at a tertiary medical center in western Pennsylvania. *Antimicrob. Agents Chemother.* **2009**, *53*, 4733–4739. [CrossRef] [PubMed]

39. Mavroidi, A.; Miriagou, V.; Malli, E.; Stefos, A.; Dalekos, G.N.; Tzouvelekis, L.S.; Petinaki, E. Emergence of *Escherichia coli* sequence type 410 (ST410) with KPC-2 β-lactamase. *Int. J. Antimicrob. Agents* **2012**, *39*, 247–250. [CrossRef] [PubMed]

40. Huber, H.; Zweifel, C.; Wittenbrink, M.M.; Stephan, R. ESBL-producing uropathogenic *Escherichia coli* isolated from dogs and cats in Switzerland. *Vet. Microbiol.* **2013**, *162*, 992–996. [CrossRef]

41. López-Cerero, L.; Egea, P.; Serrano, L.; Navarro, D.; Mora, A.; Blanco, J.; Doi, Y.; Paterson, D.L.; Rodríguez-Baño, J.; Pascual, A. Characterisation of clinical and food animal *Escherichia coli* isolates producing CTX-M-15 extended-spectrum β-lactamase belonging to ST410 phylogroup A. *Int. J. Antimicrob. Agents* **2011**, *37*, 365–367. [CrossRef]

42. Silva, K.C.; Moreno, M.; Cabrera, C.; Spira, B.; Cerdeira, L.; Lincopan, N.; Moreno, A.M. First characterization of CTX-M-15-producing *Escherichia coli* strains belonging to sequence type (ST) 410, ST224, and ST1284 from commercial swine in South America. *Antimicrob. Agents Chemother.* **2016**, *60*, 2505–2508. [CrossRef]

43. Fischer, J.; Rodríguez, I.; Baumann, B.; Guiral, E.; Beutin, L.; Schroeter, A.; Kaesbohrer, A.; Pfeifer, Y.; Helmuth, R.; Guerra, B. *bla*CTX-M-15-carrying *Escherichia coli* and *Salmonella* isolates from livestock and food in Germany. *J. Antimicrob. Chemother.* **2014**, *69*, 2951–2958. [CrossRef]

44. Schaufler, K.; Semmler, T.; Wieler, L.H.; Wöhrmann, M.; Baddam, R.; Ahmed, N.; Müller, K.; Kola, A.; Fruth, A.; Ewers, C.; et al. Clonal spread and interspecies transmission of clinically relevant ESBL-producing *Escherichia coli* of ST410—Another successful pandemic clone? *FEMS Microbiol. Ecol.* **2015**, *92*.

45. Yamaji, R.; Friedman, C.R.; Rubin, J.; Suh, J.; Thys, E.; McDermott, P.; Hung-Fan, M.; Riley, L.W. A population-based surveillance study of shared genotypes of *Escherichia coli* isolates from retail meat and suspected cases of urinary tract infections. *mSphere* **2018**, *3*, e00179-18. [CrossRef]

46. Schink, A.K.; Kadlec, K.; Kaspar, H.; Mankertz, J.; Schwarz, S. Analysis of extended-spectrum-β-lactamase-producing *Escherichia coli* isolates collected in the GERM-Vet monitoring programme. *J. Antimicrob. Chemother.* **2013**, *68*, 1741–1749. [CrossRef] [PubMed]

47. Aizawa, J.; Neuwirt, N.; Barbato, L.; Neves, P.R.; Leigue, L.; Padilha, J.; Pestana de Castro, A.F.; Gregory, L.; Lincopan, N. Identification of fluoroquinolone-resistant extended-spectrum beta-lactamase (CTX-M-8)-producing *Escherichia coli* ST224, ST2179 and ST2308 in buffalo (*Bubalus bubalis*). *J. Antimicrob. Chemother.* **2014**, *69*, 2866–2869. [CrossRef] [PubMed]

48. Guenther, S.; Aschenbrenner, K.; Stamm, I.; Bethe, A.; Semmler, T.; Stubbe, A.; Stubbe, M.; Batsajkhan, N.; Glupczynski, Y.; Wieler, L.H.; et al. Comparable high rates of extended-spectrum-beta-lactamase-producing *Escherichia coli* in birds of prey from Germany and Mongolia. *PLoS ONE* **2012**, *7*, e53039. [CrossRef] [PubMed]

49. Eklund, M.; Thomson, K.; Jalava, J.; Niiinistö, K.; Grönthal, T.; Piiparinen, H.; Rantala, M. Epidemiological comparison of extended-spectrum beta-lactamase (ESBL)-producing *Enterobacteriaceae from equine patients at the finnish veterinary teaching hospital in 2011–2014. ePoster ECMID 2015. Diagn. Bacteriol. Gen. Microbiol.* **2015**. Unpublished work.

50. Lv, L.; Cao, Y.; Yu, P.; Huang, R.; Wang, J.; Wen, Q.; Zhi, C.; Zhang, Q.; Liu, J.H. Detection of *mcr-1* gene among *Escherichia coli* isolates from farmed fish and characterization of *mcr-1* bearing IncP plasmids. *Antimicrob. Agents Chemother.* **2018**, *62*, e02378-17. [CrossRef]

51. Dierikx, C.M.; van Duijkeren, E.; Schoormans, A.H.W.; van Essen-Zandbergen, A.; Veldman, K.; Kant, A.; Huijsdens, X.W.; van der Zwaluw, K.; Wagenaar, J.A.; Mevius, D.J. Occurrence and characteristics of extended-spectrum-β-lactamase- and AmpC-producing clinical isolates derived from companion animals and horses. *J. Antimicrob. Chemother.* **2012**, *67*, 1368–1374. [CrossRef]

52. Fu, T.; Du, X.D.; Cheng, P.P.; Li, X.R.; Zhao, X.F.; Pan, Y.S. Characterization of an *rmtB*-carrying IncI1 ST136 plasmid in avian *Escherichia coli* isolates from chickens. *J. Med. Microbiol.* **2016**, *65*, 387–391. [CrossRef]

53. Yang, R.S.; Feng, Y.; Lv, X.Y.; Duan, J.H.; Chen, J.; Fang, L.X.; Xia, J.; Liao, X.P.; Sun, J.; Liu, Y.H. Emergence of NDM-5- and MCR-1-producing *Escherichia coli* clones ST648 and ST156 from a single muscovy duck (*Cairina moschata*). *Antimicrob. Agents Chemother.* **2016**, *60*, 6899–6902. [CrossRef]

54. Geue, L.; Schares, S.; Mintel, B.; Conraths, F.J.; Muller, E.; Ehricht, R. Rapid microarray-based genotyping of enterohemorrhagic *Escherichia coli* serotype O156:H25/H-/Hnt isolates from cattle and clonal relationship analysis. *Appl. Environ. Microbiol.* **2010**, *76*, 5510–5519. [CrossRef]

55. Thomsen, M.C.F.; Ahrenfeldt, J.; Cisneros, J.L.B.; Jurtz, V.; Larsen, M.V.; Hasman, H.; Aarestrup, F.M.; Lund, O. A bacterial analysis platform: An integrated system for analysing bacterial whole genome sequencing data for clinical diagnostics and surveillance. *PLoS ONE* **2016**, *11*, e0157718. [CrossRef]

56. Clermont, O.; Christenson, J.K.; Denamur, E.; Gordon, D.M. The Clermont *Escherichia coli* phylo-typing method revisited: Improvement of specificity and detection of new phylo-groups. *Environ. Microbiol. Rep.* **2013**, *5*, 58–65. [CrossRef] [PubMed]

57. Picard, B.; Garcia, J.S.; Gouriou, S.; Duriez, P.; Brahimi, N.; Bingen, E.; Elion, J.; Denamur, E. The link between phylogeny and virulence in *Escherichia coli* extraintestinal infection. *Infect. Immun.* **1999**, *67*, 546–553. [PubMed]

58. Lambrecht, E.; Van Coillie, E.; Van Meervenne, E.; Boon, N.; Heyndrickx, M.; Van de Wiele, T. Commensal *E. coli* rapidly transfer antibiotic resistance genes to human intestinal microbiota in the Mucosal Simulator of the Human Intestinal Microbial Ecosystem (M-SHIME). *Int. J. Food Microbiol.* **2019**, *311*, 108357. [CrossRef] [PubMed]

59. Chakraborty, A.; Saralaya, V.; Adhikari, P.; Shenoy, S.; Baliga, S.; Hegde, A. Characterization of Escherichia coli Phylogenetic Groups Associated with Extraintestinal Infections in South Indian Population. *Ann. Med. Health Sci. Res.* **2015**, *5*, 241–246. [PubMed]

60. Mammeri, H.; Poirel, L.; Nordmann, P. Extension of the hydrolysis spectrum of AmpC β-lactamase of *Escherichia coli* due to amino acid insertion in the H-10 helix. *J. Antimicrob. Chemother.* **2007**, *60*, 490–494. [CrossRef]

61. Miriagou, V.; Cornaglia, G.; Edelstein, M.; Galani, I.; Giske, C.G.; Gniadkowski, M.; Malamou-Lada, E.; Martinez-Martinez, L.; Navarro, F.; Nordmann, P.; et al. Acquired carbapenemases in Gram-negative bacterial pathogens: Detection and surveillance issues. *Clin. Microbiol. Infect.* **2010**, *16*, 112–122. [CrossRef]
62. Jacoby, G.A. AmpC beta-lactamases. *Clin. Microbiol. Rev.* **2009**, *22*, 161–182. [CrossRef]

Article

Antimicrobial Resistance, Virulence, and Genetic Lineages of Staphylococci from Horses Destined for Human Consumption: High Detection of *S. aureus* Isolates of Lineage ST1640 and Those Carrying the *lukPQ* Gene

Olouwafemi Mistourath Mama [1], Paula Gómez [1], Laura Ruiz-Ripa [1], Elena Gómez-Sanz [2,3], Myriam Zarazaga [1] and Carmen Torres [1,*]

[1] Departamento de Agricultura y Alimentación, Universidad de La Rioja, 26006 Logroño, Spain; femimama92@hotmail.com (O.M.M.); paula_gv83@hotmail.com (P.G.); laura_ruiz_10@hotmail.com (L.R.-R.); myriam.zarazaga@unirioja.es (M.Z.)

[2] Laboratory of Food Microbiology, Institute of Food, Nutrition and Health, ETH Zurich, 8092 Zurich, Switzerland; elena.gomez@hest.ethz.ch

[3] Área de Microbiología Molecular, Centro de Investigación Biomédica de La Rioja (CIBIR), 26006 Logroño, Spain

* Correspondence: carmen.torres@unirioja.es

Received: 1 October 2019; Accepted: 30 October 2019; Published: 1 November 2019

Simple Summary: Staphylococci are opportunistic pathogens which colonize humans and animals. Zoonotic transfer of staphylococcal species between domestic animals and humans is common and can occur through direct contact, the environment, and animal-derived food processing, implying a risk of the spread of antimicrobial resistance mechanisms and virulence factors into different ecosystems. Our work aimed at studying the diversity of staphylococcal species in nasal and faecal samples of healthy horses intended for human consumption and their resistance and virulence determinants. Staphylococci were detected in 90% and 66% of nasal and faecal samples tested, respectively. Eight staphylococcal species were detected, with the most prevalent ones being *Staphylococcus aureus* (all isolates were methicillin-susceptible), *Staphylococcus delphini*, and *Staphylococcus sciuri*. The predominant *S. aureus* lineage, ST1640, is associated with horses for the first time in this study. *S. aureus* isolates, except those of lineage ST1640, produced equid-adapted leukocidin (LukPQ) and blocker of equine complement system activation (eqSCIN). The toxic shock syndrome toxin-encoding gene was also detected in some *S. aureus* isolates. Multidrug resistance was observed among *S. sciuri* isolates, but not among *S. aureus*. Measures of hygiene and control should be implemented during horse slaughter and meat processing.

Abstract: This work aimed to determine the frequency and diversity of *Staphylococcus* species carriage in horses intended for human consumption, as well as their resistance and virulence determinants. Eighty samples (30 nasal; 50 faecal) were recovered from 73 healthy horses in a Spanish slaughterhouse. The samples were cultured for staphylococci and methicillin-resistant staphylococci (MRS) recovery. The phenotype/genotype of antimicrobial resistance was analysed for all isolates. The *spa*-type and sequence-type (ST) were determined in *Staphylococcus aureus* strains; moreover, the presence of virulence and host-adaptation genes (*tst, eta, etb, pvl, lukPQ, scn-eq*, and *scn*) was studied by PCR. *Staphylococcus* species were detected in 27/30 (90%) and 33/50 (66%) of nasal and faecal samples, respectively. Ninety isolates belonging to eight species were recovered, with predominance of *S. aureus* (*n* = 34), *Staphylococcus delphini* (*n* = 19), and *Staphylococcus sciuri* (*n* = 19). *S. aureus* strains were all methicillin-susceptible (MSSA), 28/34 were susceptible to all the antibiotics tested, and the remaining six showed resistance to (gene-detected) streptomycin (*ant* (6)-*Ia*), penicillin (*blaZ*), and trimetroprim/sulphametoxazole (SXT) (*dfrA, dfrG*). The lineage ST1640/t2559 was predominant

(n = 21). The genes *lukPQ* and *scn-eq* were present in all but the ST1640 isolates. Three *S. sciuri* isolates were multidrug-resistant. Healthy horses in Spain seem to be a reservoir for virulent MSSA and the lineage ST1640, although the presence of the latter in horses is described for the first time in this study. Moreover, the equine-adapted leukocidin gene *lukPQ* is frequent among *S. aureus* strains. A large variety of staphylococcal species with low antibiotic resistance rate were also observed.

Keywords: healthy horses; staphylococci; MSSA; ST1640; *lukPQ*

---

1. Introduction

Staphylococci are commensal bacteria that generally colonize nares, skin, and mucous membranes of humans and of wild and domestic animals, although some species are opportunistic pathogens [1–6]. Horses have been described as carriers of staphylococcal species and methicillin-resistant staphylococci (MRS) [7–10]. Coagulase-positive staphylococci (CoPS) such as *Staphylococcus aureus*, *Staphylococcus intermedius*, *Staphylococcus delphini*, and *Staphylococcus pseudintermedius* are frequently reported as colonizers or infectious agents in horses [8–12]. Coagulase-negative staphylococci (CoNS) have been described as causative agents of mastitis, wound infections, and skin abscesses in various animals, including horses [13].

Staphylococcal infections are a major issue in both human and veterinary medicine, and their role in severe diseases has increased with the acquisition of antimicrobial resistance mechanisms [11,13]. Moreover, *S. aureus* has a large variety of virulence factors, such as staphylococcal enterotoxins, toxic shock syndrome toxin (TSST-1), or leukocidins, among others [14]. Leukocidins are a family of bicomponent pore-forming toxins involved in *S. aureus* pathogenicity [15]. To date, six leukocidins have been identified, including Panton Valentine leukocidin (*lukF/lukS-PV*), LukMF', and the novel equid-adapted leukocidin LukPQ, which are related to phage-encoded genes mainly found in humans, ruminants, and equines, respectively [15,16]. LukPQ, encoded by the 45-kb prophage φSaeq1, was found to be strongly associated with *S. aureus* from horses and donkeys. This leukocidin preferentially destroys neutrophils with higher efficiency than its closest fellow, LukED [15]. It was recently revealed that the prophage φSaeq1 also encodes a novel variant of staphylococcal complement inhibitor SCIN-A (termed eqSCIN, encoded by *scn-eq*) which shares 57.8% amino acid identity with SCIN-A (encoded by *scn*) from human *S. aureus* [17]. eqSCIN is a potent blocker of equine complement system activation, which plays an important role in *S. aureus* host adaptation. Whereas SCIN-A isolates exclusively inhibit human complement, eqSCIN represents the first animal-adapted SCIN variant that functions in a broader range of hosts (horses, humans, and pigs) [17].

The presence of staphylococcal species in horses is of public health concern since the potential transfer of *Staphylococcus* spp. and their resistance and virulence genes between healthy humans and domestic animals has been evidenced [18–20]. Direct contact may be a way of transmission, but other vehicles, such as the environment and food, should be taken into consideration. In Spain, horse meat is used for human meat consumption; hence, it is important to determine the diversity of staphylococcal species colonizing the mentioned animal species. In that context, this work aimed to identify the different species of staphylococci present in nares and faeces of healthy horses destined to human consumption, as well as the antimicrobial resistance phenotype and genotype of the recovered isolates, and the virulence traits for *S. aureus* species.

2. Material and Methods

2.1. Sample Recovery

A total of 80 samples (nasal: n = 30 and faecal: n = 50) were recovered with sterile swabs from 73 healthy horses intended for human consumption and kept in Amies transport medium

(Copan, Murrieta/USA). Seven animals were tested for both types of samples. Animals came from 19 Spanish regions before they were transported to a slaughterhouse located in Northern Spain, where samples were taken in February 2012.

2.2. Staphylococcus spp. Isolation, Identification, and DNA Extraction

The samples were first inoculated in Brain Heart Infusion (BHI, supplemented with NaCl 6.5%) broth (Conda, Madrid/Spain) and incubated at 37 °C for 24 h. After growth, the bacterial culture was distributed on plates of mannitol–salt–agar (Conda, Madrid/Spain) and oxacillin resistance screening agar base (Oxoid, Hampshire/England) for staphylococci and MRS recovery, respectively. Up to four colonies/plate with staphylococcal morphology were isolated and subjected to the DNase agar test (Conda, Madrid/Spain). Identification was performed by PCR (for CoPS isolates) [21] and by matrix-assisted laser desorption/ionization time of flight (MALDI-TOF) mass spectrometry (Bruker, MA, USA) (for all the staphylococci).

The DNA extraction was performed as follows: one colony was resuspended in 45 μL of milli-Q water and 5 μL of lysostaphin (1 mg/mL). The suspension was warmed in a water bath at 37 °C during 10 min. Then 45 μL of milli-Q water, 150 μL of Tris (0.1 M, pH 8.5), and 5 μL of proteinase K (2 mg/mL) were added to the suspension before it was heated again in a water bath at 60 °C for 10 min and at 100 °C for 5 min. Finally, centrifugation was performed at 12,000 rpm for 3 min and the supernatant was kept for further experiences.

2.3. Antimicrobial Susceptibility and Resistance Genes

Susceptibility testing to penicillin, cefoxitin, gentamicin, tobramycin, tetracycline, erythromycin, clindamycin, chloramphenicol, ciprofloxacin, linezolid, and trimethoprim/sulfamethoxazole (SXT) was performed by disk-diffusion method according to the Clinical Laboratory Standards Institute recommendations [22]. Susceptibility to streptomycin was also tested (CASFM 2018). The presence of the following antimicrobial resistance genes was determined by PCR, in accordance with the identified resistance phenotypes: beta-lactams (*mecA*, *blaZ*), tetracycline (*tet*(K), *tet*(L), and *tet*(M)), macrolides-lincosamides (*erm*(A), *erm*(B), *erm*(C), *erm*(T), *msr*(A), *lnu*(A), *lnu*(B), and *vgaA*), streptomycin (*str* and *ant*(6)-Ia), chloramphenicol (*fexA* and *fexB*), and SXT (*dfrA*, *dfrD*, *dfrG* and *dfrK*) [23–27].

2.4. Molecular Typing

For the *S. aureus* isolates, *spa*-typing was performed as previously described [28]. Multilocus sequence typing (MLST) was determined for representative isolates (one isolate of each *spa*-type, except for the *spa*-type t2559 for which three associated isolates were chosen). For this purpose, PCR and sequencing of seven housekeeping genes (www.pubmlst.org) were performed to define the sequence type (ST) and the clonal complex (CC). Additionally, detection of *agr* allotypes was carried out by two multiplex PCRs in all isolates [29]. For *S. delphini* isolates, a PCR with specific primers was performed to classify them in two groups (A or B) as previously described [30].

2.5. Virulence Genes

For *S. aureus* isolates, the presence of the genes encoding the toxic shock syndrome toxin (*tst*) and exfoliative toxin A (*eta*) and B (*etb*) was studied by PCR [26]. In addition, the genes encoding the leukocidins of Panton-Valentine (*lukF/lukS*-PV) and LukPQ were studied by PCR and sequencing [15,26]. The presence of *scn* (gene for SCIN-A) [31] and *scn-eq* (gene for eqSCIN) was analysed by PCR and sequencing. The *scn-eq* PCR was performed using a pair of primers designed in this study (eqSCN-F: TGCTGCTTTTGCTTTGTATCC and eqSCN-R: TGCAGGAGTTTTAGTTGCAGTTTT) and the following conditions: 94 °C for 3 min, followed by 30 cycles of 1 min at 94 °C, 2 min at 61.5 °C, and 3 min at 72 °C, with a final extension at 72 °C for 10 min.

3. Results

Staphylococcus isolates were detected in 27/30 (90%) of nasal samples and in 33/50 (66%) of faecal samples. The seven animals tested for both types of samples were positive for staphylococcal species in both cases.

A total of 90 isolates of eight species were detected in the positive nasal/faecal samples: *S. aureus* (*n* = 34), *S. delphini* (*n* = 19), *Staphylococcus sciuri* (*n* = 19), *Staphylococcus simulans* (*n* = 4), *Staphylococcus fleurettii* (*n* = 2), *Staphylococcus lentus* (*n* = 2), *Staphylococcus saprophyticus* (*n* = 2), *Staphylococcus xylosus* (*n* = 2), *Staphylococcus haemolyticus* (*n* = 2), *Staphylococcus schleiferi* (*n* = 2), *Staphylococcus vitulinus* (*n* = 1), and *Staphylococcus hyicus* (*n* = 1) (Tables 1 and 2). Twenty-four samples harboured at least two isolates of either distinct species, distinct antibiotic resistance phenotypes or different *spa*-types.

3.1. S. aureus Isolates: Molecular Characteristics, Antimicrobial Resistance, and Virulence Determinants

S. aureus was detected in 17/30 (56.6%) and 16/50 (32%) of the nasal and faecal samples, respectively. The 34 isolates recovered (nasal origin: *n* = 18; faecal origin: *n* = 16) were ascribed to five *spa*-types (t2559, t3269, t127, t1294 and t549), four STs (ST1640, ST1, ST816 and ST1660), and four *agr*-types (I, II, III and IV) (Table 1). One isolate per sample was detected except for one nasal sample which harboured two *S. aureus* isolates of different *spa*-types (t2559 and t1294). Most of the isolates were susceptible to all the antimicrobials tested (*n* = 28; 82.4%), while the remaining six showed the following resistance phenotypes (number of isolates; genes detected): streptomycin (3; *ant*(6)-*Ia*), penicillin+streptomycin (1; *blaZ*, *ant*(6)-*Ia*, *str*), and penicillin + SXT (2; *blaZ*, *dfrA*, *dfrG*). In addition, three nasal isolates (all ST816) hosted the gene *tst*. Moreover, all isolates but those of the lineage ST1640 harboured the genes *lukPQ* and *scn-eq*. The isolates of lineage ST1640 were collected from animals which came from six regions, most of them of Southern Spain.

The major *S. aureus* lineage in horses found in this study was the ST1640 associated to nine and 12 strains of nasal and faecal origins, respectively. In addition, the antimicrobial resistance rates, phenotypes and genotypes are similar in isolates of both origins (Table 1). Concerning the virulence genes, the isolates of the lineage ST1640 were the only ones which lacked the LukPQ determinants. None of the isolates of this study harboured the *scn* gene, a marker of the human immune evasion cluster (IEC).

Table 1. Characteristics of 34 *Staphylococcus aureus* isolates recovered from nasal and faecal samples of horses at a slaughterhouse.

Sample Source	Number of Strains	ST/CC [a] (Number of Strains)	*Spa*-type (Number of Strains)	*Agr-Type*	Antimicrobial Resistance		Virulence Genes (Number of Strains)
					Phenotype [b] (Number of Strains)	Genotype (Number of Strains)	
Nasal	18	ST1640 (9)	t2559 (9)	IV	STR (1)	*ant(6)-Ia*	-
					SUSCEPTIBLE (8)	-	-
		ST1/CC1 (5)	t3269 (3)	III	SUSCEPTIBLE (3)	-	*lukPQ* (3), *scn-eq* (3)
			t127 (2)	III	PEN, SXT (1)	*blaZ, dfrG*	*lukPQ, scn-eq*
					SUSCEPTIBLE (1)	-	*lukPQ, scn-eq*
		ST816/CC479 (3)	t1294 (3)	II	SUSCEPTIBLE (3)	-	*tst* (3), *lukPQ* (3), *scn-eq* (3)
		ST1660/CC9 (1)	t549 (1)	II	PEN, STR	*blaZ, ant(6)-Ia, str*	*lukPQ, scn-eq*
Faecal	16	ST1640 (12)	t2559 (12)	IV	STR (1)	-	-
					SUSCEPTIBLE (11)	-	-
		ST1/CC1 (3)	t127 (2)	III	PEN, SXT (1)	*blaZ, dfrA, dfrG*	*lukPQ, scn-eq*
			t386 (1)	III	SUSCEPTIBLE (1)	-	*lukPQ, scn-eq*
					STR	-	*lukPQ, scn-eq*
		ST133/CC133 (1)	t2420 (1)	I	SUSCEPTIBLE	-	*lukPQ, scn-eq*

[a] The multilocus sequence typing (MLST) was performed for one isolate of each *spa*-type, and for three isolates of *spa*-type t2559; [b] PEN: penicillin; STR: streptomycin; SXT: trimethoprim/sulphametoxazole.

Table 2. Characteristics of 56 non-*aureus* staphylococcal isolates recovered from nasal and faecal samples of horses at the slaughterhouse.

Sample Source	(Number of Strains)	Species (Number of Strains)	Antimicrobial Resistance	
			Phenotype [c] (Number of Strains)	Genotype (Number of Strains)
Nasal	30	*S. delphini* [a] (11)	SUSCEPTIBLE (11)	-
		S. sciuri (10)	STR (1)	*str (1)*
			PEN, FOX (2)	*mecA* (2)
			PEN, FOX, TET (2)	*mecA* (2), *tet*(K) (2), *tet*(L) (2)
			PEN, FOX, CLI (1)	*mecA, lnu*(A) (1)
			PEN, FOX, STR, TET[e] (1)	*mecA, str, tet*(K), *tet*(L) (1)
			SUSCEPTIBLE (3)	-
		S. fleurettii (2)	SUSCEPTIBLE (2)	-
		S. lentus (2)	ERY, CLI [d] (1)	*erm*(A), *erm*(B), *msr*(A) (1)
			PEN, FOX, STR (1)	*mecA, str* (1)
		S. saprophyticus (2)	SUSCEPTIBLE (2)	-
		S. xylosus (2)	CLI (1)	-
			SUSCEPTIBLE (1)	-
		S. haemolyticus (1)	SUSCEPTIBLE (1)	-
Faecal	26	*S. delphini* [b] (8)	SUSCEPTIBLE (8)	-
		S. sciuri (9)	PEN, FOX (6)	*mecA* (6)
			PEN, FOX, STR, TET [e] (1)	*mecA, str, tet*(K), *tet*(L) (1)
			ERY, CLI [d], STR, CHL [e] (1)	*erm*(C), *str, fexA* (1)
			SUSCEPTIBLE (1)	-
		S. simulans (4)	SUSCEPTIBLE (4)	-
		S. schleiferi (2)	SUSCEPTIBLE (2)	-
		S. haemolyticus (1)	SUSCEPTIBLE (1)	-
		S. vitulinus (1)	SUSCEPTIBLE (1)	-
		S. hyicus (1)	SUSCEPTIBLE (1)	-

[a] type B: *n* = 10; type A: *n* = 1; [b] type B: *n* = 7; type A: *n* = 1; [c] PEN: penicillin, FOX: cefoxitin, ERY: erythromycin, CLI: clindamycin, TET: tetracycline, STR: streptomycin, CHL: chloramphenicol; [d] inducible resistance phenotype. [e] multidrug resistance phenotype.

3.2. Non-aureus Staphylococcus Species: Molecular Characteristics, and Antimicrobial Resistance

Among the other species identified, one belonged to *Staphylococcus intermedius* group (SIG), *S. delphini*, and was present in 11 (36.6%) of the nasal samples and in eight (16%) of the faecal samples. Nineteen *S. delphini* isolates were detected in total (type B: n = 17; type A: n = 2). All isolates were susceptible to the antimicrobials tested (Table 2).

The 37 remaining isolates belonged to coagulase-negative staphylococci (CoNS) group. CoNS were present in 16 (53.3%) of the nasal samples and 15 (30%) of the faecal samples. One isolate per sample was recovered except for five samples which harboured two or three isolates of distinct species. Resistance to at least one antimicrobial agent was detected in 52.6% and 44.4% of the nasal and faecal isolates, respectively. Seventy percent of the nasal resistant isolates and 100% of the faecal resistant isolates belonged to the predominant species *S. sciuri*. The other species with resistant isolates were *S. lentus* and *S. xylosus*. Moreover, three *S. sciuri* isolates showed a multidrug resistance (MDR) phenotype, meaning that they were resistant to one agent in three or more antimicrobial categories (Table 2). Globally, the following rates, phenotypes, and genotypes of antimicrobial resistance were reported among CoNS isolated from horses (detection rate; resistance genes detected): penicillin and cefoxitin (37.8%; *mecA*), erythromycin (5.4%; *erm* (A), *erm* (B), *erm* (C) or *msr* (A)), clindamycin (10.8%; *lnu* (A)), tetracycline (10.8%; *tet* (K) and *tet* (L)), streptomycin (13.5%; *str*), and chloramphenicol (2.7%; *fexA*).

3.3. Comparison of Nasal and Faecal Samples of Seven Healthy Horses

Seven animals could be tested for both nasal and faecal samples. The results are displayed in Table 3. All the animals that carried *S. aureus* in their nostrils also had this microorganism in their faeces (n = 5). In 3/5 cases, the *S. aureus* isolates belonged to the same genetic lineage ST1640. More than one staphylococcal species was detected in five of seven nasal samples, while faecal samples predominantly carried a single staphylococcal species (*S. aureus* in 5/7 cases). The lineage ST1640 was predominant in both nasal and faecal samples.

Table 3. Comparison of staphylococci recovered from nasal and faecal samples from seven healthy horses.

Animal	Nasal Samples			Faecal Samples		
Animal Code	Species Detected (Number of Strains)	Type A/B or *spa*-type /ST	Antimicrobial Resistance Phenotype [a]	Species Detected (Number of Strains)	Type A/B or *spa*-Type/ST	Antimicrobial Resistance Phenotype [a]
1	*S. delphini* (1)	Type B	SUSCEPTIBLE	*S. simulans* (1)	-	SUSCEPTIBLE
	S. haemolyticus (1)	-	SUSCEPTIBLE			
25	*S. aureus* (1)	t2559/ST1640	SUSCEPTIBLE	*S. aureus* (1)	t2420/ST133	SUSCEPTIBLE
26	*S. aureus* (1)	t2559/ST1640	SUSCEPTIBLE	*S. aureus* (1)	t2559/ST1640	SUSCEPTIBLE
	S. sciuri (1)	-	PEN, FOX, STR, TET			
27	*S. aureus* (1)	t549/ST1660	PEN, STR	*S. aureus* (1)	t127/ST1	PEN, SXT
	S. delphini (1)	Type A	SUSCEPTIBLE	*S. delphini* (1)	Type B	SUSCEPTIBLE
28	*S. aureus* (1)	t2559/ST1640	SUSCEPTIBLE	-	-	-
29	*S. aureus* (1)	t2559/ST1640	SUSCEPTIBLE	*S. aureus* (1)	t2559/ST1640	SUSCEPTIBLE
	lentus (1)	-	PEN, FOX, STR			
30	*S. aureus* (1)	t2559/ST1640	SUSCEPTIBLE	*S. aureus* (1)	t2559/ST1640	SUSCEPTIBLE
	S. sciuri (1)	-	SUSCEPTIBLE			

[a] PEN: penicillin; FOX: cefoxitin; TET: tetracycline; STR: streptomycin; SXT: trimetroprim-sulphametoxazole.

4. Discussion

High *S. aureus* occurrence has been detected among both nasal and faecal samples of healthy horses destined for human consumption (56.6% and 32%, respectively). According to previous works on healthy horses from various farms in Germany and Denmark, the occurrence of *S. aureus* in nasal samples was much lower (6.7% and 13.5%, respectively) [32,33]. Alternatively, a recent Italian study showed that the prevalence of methicillin resistant *S. aureus* (MRSA) in horses tested in slaughterhouses (7%) was significantly higher than those tested on farms and racecourses [34]. In our study, however, no MRSA was detected among the population tested. Islam and collaborators observed that 63.3% of the *S. aureus* strains recovered were MSSA strains, mostly assigned to ST1/t127 and ST1660/t549 [33]. However, the lineages ST1, ST1660, and ST133 are also frequent among MSSA from horses [32,33,35]. Our strains were mostly associated to ST1640/t2559 (*n* = 21), although ST1 (*n* = 8), ST1660 (*n* = 1), and ST133 were also detected. To our knowledge, the lineage ST1640/t2559 is here detected for the first time among horse samples. Nonetheless, the *spa*-type t2559 was previously found (associated to CC5/CC30) in the nostrils of patients of general practitioners with no sign of infections in the Netherlands [36]. The lack of the *scn* gene suggests that none of the strains were of human origin.

Interestingly, all strains but those of the lineage ST1640/t2559 harboured the equine-adapted leukocidin determinant *lukPQ* (prevalence of 38%) and the *scn-eq* gene. These findings suggest that the ST1640 might have jumped recently from another source to the equine environment. On the other hand, an international equid collection study reported *lukPQ* values ranging from 0% to 50%, indicating either (1) the absence of these genes may also be a common feature among horse isolates or, again, (2) a reflection of an early phase of those isolates in the adaptation to this host [15]. Otherwise, it was revealed that *lukPQ* and *scn-eq*, both encoded by the prophage φSaeq1, are prone to occur together and were associated with the clonal complexes CC1, CC133, CC1660, CC350, and CC522 [15,17]. These findings are confirmed by our results (*lukPQ* and *scn-eq* genes associated with CC1, CC133, CC1660, and ST816). The phage-encoded leukocidin LukPQ displays a high toxicity towards equine neutrophils, while the eqSCIN blocks complement activity in equine serum, which implies an important role in the evasion of *S. aureus* of the equid host defence mechanism [15,17]. Moreover, LukPQ has a broad host range as at high concentrations it is capable of lysing bovine and to some extent human neutrophils. Its transmission to human *S. aureus* strains could enhance its pathogenicity. The toxic shock syndrome gene *tst* was detected in strains of ST816, even though *tst* is generally observed among small ruminant isolates [4,35,37]. The presence of these virulence factors in healthy horses destined for human consumption might be of concern for food security and public health since it can spread through handling and processing.

Regarding the antimicrobial resistance, a low prevalence of resistant strains among *S. aureus* isolates was observed (17.6%, *n* = 4). They showed resistance to penicillin, streptomycin, and SXT, which are antibiotics frequently used in veterinary medicine [38].

Other staphylococcal species were identified from the horses studied, with predominance of *S. delphini* and *S. sciuri*. *S. delphini* is described as a colonizer of a wide variety of animal species (Equidae, Mustelidae, dolphins, pigeons, cinerous vulture, among others) [9,39–41]. Here, the *S. delphini* group B revealed a predominance in horses. A similar trend was observed by Stull and collaborators in Canada [9], as well as in wild birds in Spain [40]. The high susceptibility to the antibiotics observed among our strains was in accordance with previous results in donkeys and might be due to a lower selective pressure exerted on these animal species [9,41]. Unfortunately, data on antimicrobial therapy or exposure level of these animals were not available.

S. sciuri was the predominant species with resistance to methicillin, as previously reported among equine staphylococcal isolates [42]. This species hosts a native *mecA* homologue (*mecA1*) estimated to be the origin of the *mecA* gene for MRS [42]. Three of our strains showed an MDR phenotype. Those resistance genes could be disseminated among horses and humans through contact and derived food manipulation, which would be a risk for animals and human health. In fact, CoNS and methicillin resistant CoNS of the species detected in this study (*S. epidermidis*, *S. haemolyicus*, *S. sciuri*, *S. xylosus*, or *S vitulinus* among others) are frequently isolated from healthy and infected horses and, sometimes, among equine personnel [13,32,42,43].

On the other hand, comparison of staphylococcal carriage between the nostrils and the faeces among several animals in this study indicate a higher carriage rate in nasal samples. These results are in agreement with former data, which describe human and animal skin and mucosa, especially the nares, as the most frequent carriage site for staphylococci [3,44]. Remarkably, our results indicate that the gut and nasal microbiota of these animals is similar when referring to staphylococcal species.

5. Conclusions

This study provides data on the staphylococcal carriage of healthy horses. A high prevalence of MSSA, mostly susceptible to the antibiotics tested but carrying important virulence genes (*lukPQ*, *scn-eq*, and *tst*), is highlighted. The detection and predominance of the *S. aureus* lineage ST1640 in horses is noteworthy, as it represents its first description in horses. Furthermore, a high diversity of species among non-*S. aureus* isolates was observed, including CoPS and MRCoNS. Due to current evidence on the influence of animal-derived food in the dissemination of staphylococci and their resistance and virulence genes, strict measures of hygiene and control must be taken for horses at slaughter and for meat processing.

Author Contributions: Conceptualization, C.T.; methodology and validation: C.T. and M.Z.; formal analysis, O.M.M.; investigation, O.M.M., P.G., and L.R.-R.; data curation, O.M.M.; writing—original draft preparation, O.M.M.; writing—review and editing, O.M.M., P.G., L.R.-R., E.G.-S., C.T., and M.Z.; visualization, O.M.M., P.G., E.G.-S., C.T., and M.Z.; supervision, C.T.; project administration, C.T.; funding acquisition, C.T. and M.Z.

Funding: This research was funded by the "Agencia Estatal de Investigación (AEI) of Spain and the Fondo Europeo de Desarrollo Regional (FEDER) of EU", project number "SAF2016-76571-R". The APC was funded by project SAF2016-76571-R of the AEI of Spain and FEDER of EU".

Acknowledgments: We acknowledge the foundation "Mujeres por África" and the "Universidad de La Rioja" (Spain) for OMM predoctoral fellowship, the "Universidad de La Rioja" (Spain) for LRR's predoctoral FPI fellowship.

Conflicts of Interest: The authors declare that they have no conflicts of interests.

References

1. von Eiff, C.; Peters, G.; Heilmann, C. Review Pathogenesis of infections due to coagulase- negative staphylococci. *Lancet Infect. Dis.* **2002**, *2*, 677–685. [CrossRef]

2. Walther, B.; Tedin, K.; Lübke-Becker, A. Multidrug-resistant opportunistic pathogens challenging veterinary infection control. *Vet. Microbiol.* **2017**, *200*, 71–78. [CrossRef] [PubMed]

3. Kluytmans, J.A.J.W. Methicillin-resistant *Staphylococcus aureus* in food products: Cause for concern or case for complacency? *Clin. Microbiol. Infect.* **2010**, *16*, 11–15. [CrossRef] [PubMed]

4. Mama, O.M.; Gómez-Sanz, E.; Ruiz-Ripa, L.; Gómez, P.; Torres, C. Diversity of staphylococcal species in food producing animals in Spain, with detection of PVL-positive MRSA ST8 (USA300). *Vet. Microbiol.* **2019**, *233*, 5–10. [CrossRef] [PubMed]

5. Mama, O.M.; Ruiz-Ripa, L.; Fernández-Fernández, R.; González-Barrio, D.; Ruiz-Fons, J.F.; Torres, C. High frequency of coagulase-positive staphylococci carriage in healthy wild boar with detection of MRSA of lineage ST398-t011. *FEMS Microbiol. Lett.* **2019**, *366*, fny292. [CrossRef] [PubMed]

6. Mama, O.M.; Ruiz-Ripa, L.; Lozano, C.; González-Barrio, D.; Ruiz-Fons, J.F.; Torres, C. High diversity of coagulase negative staphylococci species in wild boars, with low antimicrobial resistance rates but detection of relevant resistance genes. *Comp. Immunol. Microbiol. Infect. Dis.* **2019**, *64*, 125–129. [CrossRef]

7. Aslantas, Ö.; Türkyilmaz, S.; Yilmaz, M.A.; Erdem, Z.; Demir, C. Isolation and molecular characterization of Methicillin-Resistant Staphylococci from horses, personnel and environmental sites at an equine hospital in Turkey. *J. Vet. Med. Sci.* **2012**, *74*, 1583–1588. [CrossRef]

8. Burton, S.; Reid-Smith, R.; McClure, J.T.; Weese, J.S. *Staphylococcus aureus* colonization in healthy horses in Atlantic Canada. *Can. Vet. J.* **2008**, *49*, 797–799.

9. Stull, J.W.; Slavić, D.; Rousseau, J.; Scott Weese, J. *Staphylococcus delphini* and Methicillin-Resistant *S. pseudintermedius* in horses, Canada. *Emerg. Infect. Dis.* **2014**, *20*, 485–487. [CrossRef]

10. Tirosh-Levy, S.; Steinman, A.; Carmeli, Y.; Klement, E.; Navon-Venezia, S. Prevalence and risk factors for colonization with methicillin resistant *Staphylococcus aureus* and other Staphylococci species in hospitalized and farm horses in Israel. *Prev. Vet. Med.* **2015**, *122*, 135–144. [CrossRef]

11. Oguttu, J.W.; Qekwana, D.N.; Odoi, A. An Exploratory Descriptive study of antimicrobial resistance patterns of *Staphylococcus* Spp. Isolated from horses presented at a veterinary teaching hospital. *BMC Vet. Res.* **2017**, *13*, 269. [CrossRef] [PubMed]

12. Gómez-Sanz, E.; Simón, C.; Ortega, C.; Gómez, P.; Lozano, C.; Zarazaga, M.; Torres, C. First detection of methicillin-resistant *Staphylococcus aureus* ST398 and *Staphylococcus pseudintermedius* ST68 from hospitalized equines in Spain. *Zoonoses Public Health* **2014**, *61*, 192–201. [CrossRef] [PubMed]

13. Moodley, A.; Guardabassi, L. Clonal spread of methicillin-resistant coagulase-negative staphylococci among horses, personnel and environmental sites at equine facilities. *Vet. Microbiol.* **2009**, *137*, 397–401. [CrossRef] [PubMed]

14. Jans, C.; Merz, A.; Johler, S.; Younan, M.; Tanner, S.A.; Kaindi, D.W.M.; Wangoh, J.; Bonfoh, B.; Meile, L.; Tasara, T. East and West African milk products are reservoirs for human and livestock-associated *Staphylococcus aureus*. *Food Microbiol.* **2017**, *65*, 64–73. [CrossRef]

15. Koop, G.; Vrieling, M.; Storisteanu, D.M.L.; Lok, L.S.C.; Monie, T.; Van Wigcheren, G.; Raisen, C.; Ba, X.; Gleadall, N.; Hadjirin, N.; et al. Identification of LukPQ, a novel, equid-adapted leukocidin of Staphylococcus aureus. *Sci. Rep.* **2017**, *7*, e40660. [CrossRef] [PubMed]

16. McCarthy, A.J.; Lindsay, J.A. *Staphylococcus aureus* innate immune evasion is lineage-specific: A bioinfomatics study. *Infect. Genet. Evol.* **2013**, *19*, 7–14. [CrossRef]

17. De Jong, N.W.M.; Vrieling, M.; Garcia, B.L.; Koop, G.; Brettmann, M.; Aerts, P.C.; Ruyken, M.; Van Strijp, J.A.G.; Holmes, M.; Harrison, E.M.; et al. Identification of a staphylococcal complement inhibitor with broad host specificity in equid *Staphylococcus aureus* strains. *J. Biol. Chem.* **2018**, *293*, 4468–4477. [CrossRef]

18. Gómez-Sanz, E.; Torres, C.; Lozano, C.; Zarazaga, M. High diversity of *Staphylococcus aureus* and *Staphylococcus pseudintermedius* lineages and toxigenic traits in healthy pet-owning household members. Underestimating normal household contact? *Comp. Immunol. Microbiol. Infect. Dis.* **2013**, *36*, 83–94.

19. Benito, D.; Gómez, P.; Aspiroz, C.; Zarazaga, M.; Lozano, C.; Torres, C. Molecular characterization of *Staphylococcus aureus* isolated from humans related to a livestock farm in Spain, with detection of MRSA-CC130 carrying *mecC* gene: A zoonotic case? *Enferm. Infecc. Microbiol. Clin.* **2015**, *34*, 280–285. [CrossRef]

20. Velasco, V.; Buyukcangaz, E.; Sherwood, J.S.; Stepan, R.M.; Koslofsky, R.J.; Logue, C.M. Characterization of *Staphylococcus aureus* from humans and a comparison with isolates of animal origin, in North Dakota, United States. *PLoS ONE* **2015**, *10*, e0140497. [CrossRef]

21. Sasaki, T.; Tsubakishita, S.; Tanaka, Y.; Sakusabe, A.; Ohtsuka, M.; Hirotaki, S.; Kawakami, T.; Fukata, T.; Hiramatsu, K. Multiplex-PCR method for species identification of coagulase-positive staphylococci. *J. Clin. Microbiol.* **2010**, *48*, 765–769. [CrossRef] [PubMed]

22. Wayne, P.A. *Performance Standards for Antimicrobial Susceptibility Testing*, 28th ed.; Clinical and Laboratory Standard Institutes: Wayne, PA, USA, 2018.

23. Kehrenberg, C.; Schwarz, S. Distribution of florfenicol resistance genes *fexA* and *cfr* among chloramphenicol-resistant *Staphylococcus* isolates. *Antimicrob. Agents Chemother.* **2006**, *50*, 1156–1163. [CrossRef] [PubMed]

24. Liu, H.; Wang, Y.; Wu, C.; Schwarz, S.; Shen, Z.; Jeon, B.; Ding, S.; Zhang, Q.; Shen, J. A novel phenicol exporter gene, *fexB*, found in enterococci of animal origin. *J. Antimicrob. Chemother.* **2012**, *67*, 322–325. [CrossRef] [PubMed]

25. Schnellmann, C.; Gerber, V.; Rossano, A.; Jaquier, V.; Panchaud, Y.; Doherr, M.G.; Thomann, A.; Straub, R.; Perreten, V. Presence of new *mecA* and *mph*(C) variants conferring antibiotic resistance in *Staphylococcus* spp. isolated from the skin of horses before and after clinic admission. *J. Clin. Microbiol.* **2006**, *44*, 4444–4454. [CrossRef] [PubMed]

26. Benito, D.; Lozano, C.; Rezusta, A.; Ferrer, I.; Vasquez, M.A.; Ceballos, S.; Zarazaga, M.; Revillo, M.J.; Torres, C. Characterization of tetracycline and methicillin resistant *Staphylococcus aureus* strains in a Spanish

hospital: Is livestock-contact a risk factor in infections caused by MRSA CC398? *Int. J. Med. Microbiol.* **2014**, *304*, 1226–1232. [CrossRef] [PubMed]

27. Hauschild, T.; Vuković, D.; Dakić, I.; Ježek, P.; Djukić, S.; Dimitrijević, V.; Stepanović, S.; Schwarz, S. Aminoglycoside resistance in members of the *Staphylococcus sciuri* Group. *Microb. Drug Resist.* **2007**, *13*, 77–84. [CrossRef] [PubMed]

28. Shopsin, B.; Gomez, M.; Montgomery, S.O.; Smith, D.H.; Waddington, M.; Dodge, D.E.; Bost, D.A.; Riehman, M.; Naidich, S.; Kreiswirth, B.N. Evaluation of protein A gene polymorphic region DNA sequencing for typing of *Staphylococcus aureus* strains. *J. Clin. Microbiol.* **1999**, *37*, 3556–3563.

29. Shopsin, B.; Mathema, B.; Alcabes, P.; Said-Salim, B.; Lina, G.; Matsuka, A.; Martinez, J.; Kreiswirth, B.N. Prevalence of *agr* specificity groups among *Staphylococcus* aureus strains colonizing children and their guardians. *J. Clin. Microbiol.* **2003**, *41*, 456–459. [CrossRef]

30. Duran, N.; Ozer, B.; Duran, G.G.; Onlen, Y.; Demir, C. Antibiotic resistance genes & susceptibility patterns in staphylococci. *Indian J. Med. Res.* **2012**, *135*, 389–396.

31. Van Wamel, W.J.B.; Rooijakkers, S.H.M.; Van Kessel, K.P.M.; Van Strijp, J.A.G.; Ruyken, M. The innate immune modulators Staphylococcal complement inhibitor and chemotaxis inhibitory protein of *Staphylococcus aureus* are located on β-Hemolysin-converting bacteriophages. *J. Bacteriol.* **2006**, *188*, 1310–1315. [CrossRef]

32. Kaspar, U.; von Lützau, K.; Schlattmann, A.; Rösler, U.; Köck, R.; Becker, K. Zoonotic multidrug-resistant microorganisms among non-hospitalized horses from Germany. *One Health* **2019**, *7*, e100091. [CrossRef] [PubMed]

33. Islam, M.Z.; Espinosa-Gongora, C.; Damborg, P.; Sieber, R.N.; Munk, R.; Husted, L.; Moodley, A.; Skov, R.; Larsen, J.; Guardabassi, L. Horses in Denmark are a reservoir of diverse clones of methicillin-resistant and -susceptible *Staphylococcus aureus*. *Front. Microbiol.* **2017**, *8*, e543. [CrossRef] [PubMed]

34. Parisi, A.; Caruso, M.; Normanno, G.; Latorre, L.; Miccolupo, A.; Fraccalvieri, R.; Intini, F.; Manginelli, T.; Santagada, G. High Occurrence of Methicillin-Resistant *Staphylococcus aureus* in horses at slaughterhouses compared with those for recreational activities: A professional and food safety concern? *Foodborne Pathog. Dis.* **2017**, *14*, 735–741. [CrossRef] [PubMed]

35. Agabou, A.; Ouchenane, Z.; Essebe, C.N.; Khemissi, S.; Tedj, M.; Chehboub, E.; Chehboub, I.B.; Sotto, A.; Dunyach-remy, C.; Lavigne, J. Emergence of nasal carriage of ST80 and ST152 PVL+ *Staphylococcus aureus* isolates from livestock in Algeria. *Toxins (Basel)* **2017**, *9*, 303. [CrossRef]

36. Donker, G.A.; Deurenberg, R.H.; Driessen, C.; Sebastian, S.; Nys, S.; Stobberingh, E.E. The population structure of *Staphylococcus aureus* among general practice patients from The Netherlands. *Clin. Microbiol. Infect.* **2009**, *15*, 137–143. [CrossRef] [PubMed]

37. Ben Said, M.; Abbassi, M.S.; Gómez, P.; Ruiz-Ripa, L.; Sghaier, S.; El Fekih, O.; Hassen, A.; Torres, C. Genetic characterization of *Staphylococcus aureus* isolated from nasal samples of healthy ewes in Tunisia. High prevalence of CC130 and CC522 lineages. *Comp. Immunol. Microbiol. Infect. Dis.* **2017**, *51*, 37–40. [CrossRef]

38. Prestinaci, F.; Pezzotti, P.; Pantosti, A. Antimicrobial resistance: A global multifaceted phenomenon. *Pathog. Glob. Health* **2015**, *109*, 309–318. [CrossRef]

39. Guardabassi, L.; Schmidt, K.R.; Petersen, T.S.; Espinosa-Gongora, C.; Moodley, A.; Agersø, Y.; Olsen, J.E. Mustelidae are natural hosts of *Staphylococcus delphini* group A. *Vet. Microbiol.* **2012**, *159*, 351–353. [CrossRef]

40. Ruiz-Ripa, L.; Gómez, P.; Alonso, C.A.; Camacho, M.C.; de la Puente, J.; Fernández-Fernández, R.; Ramiro, Y.; Quevedo, M.A.; Blanco, J.M.; Zarazaga, M.; et al. Detection of MRSA of lineages CC130-mecC and CC398-mecA and *Staphylococcus delphini-lnu*(A) in magpies and cinereous Vultures in Spain. *Microb. Ecol.* **2019**, *78*, 409–415. [CrossRef]

41. Gharsa, H.; Slama, K.B.; Gómez-Sanz, E.; Gómez, P.; Klibi, N.; Zarazaga, M.; Boudabous, A.; Torres, C. Characterisation of nasal *Staphylococcus delphini* and *Staphylococcus pseudintermedius* isolates from healthy donkeys in Tunisia. *Equine Vet. J.* **2015**, *47*, 463–466. [CrossRef]

42. Busscher, J.F.; Van Duijkeren, E.; Sloet Van Oldruitenborgh-Oosterbaan, M.M. The prevalence of methicillin-resistant staphylococci in healthy horses in the Netherlands. *Vet. Microbiol.* **2006**, *113*, 131–136. [CrossRef] [PubMed]

43. Kern, A.; Perreten, V. Clinical and molecular features of methicillin-resistant, coagulase-negative staphylococci of pets and horses. *J. Antimicrob. Chemother.* **2013**, *68*, 1256–1266. [CrossRef] [PubMed]

44. Sakr, A.; Brégeon, F.; Mège, J.L.; Rolain, J.M.; Blin, O. Staphylococcus aureus nasal colonization: An update on mechanisms, epidemiology, risk factors, and subsequent infections. *Front. Microbiol.* **2018**, *9*, 2419. [CrossRef] [PubMed]

Article

Extended-Spectrum β-Lactamase-Producing Enterobacteriaceae in Hospitalized Neonatal Foals: Prevalence, Risk Factors for Shedding and Association with Infection

Anat Shnaiderman-Torban [1], Yossi Paitan [2,3], Haia Arielly [3], Kira Kondratyeva [4], Sharon Tirosh-Levy [1], Gila Abells-Sutton [1], Shiri Navon-Venezia [4] and Amir Steinman [1,*]

[1] Koret School of Veterinary Medicine (KSVM), The Robert H. Smith Faculty of Agriculture, Food and Environment, The Hebrew University of Jerusalem, Rehovot 761001, Israel
[2] Department of Clinical Microbiology and Immunology, Sackler Faculty of Medicine, Tel Aviv University, Tel Aviv 6997801, Israel
[3] Clinical Microbiology Lab, Meir Medical Center, Kfar Saba 4428164, Israel
[4] Department of Molecular Biology, Faculty of Natural Science, Ariel University, Ariel 40700, Israel
* Correspondence: amirst@savion.huji.ac.il

Received: 24 July 2019; Accepted: 20 August 2019; Published: 23 August 2019

Simple Summary: Multidrug-resistant (MDR) *Enterobacteriaceae* are becoming a major worldwide concern in human and veterinary medicine, mainly due to the production of extended-spectrum β-lactamases (ESBLs). These bacteria have been investigated in adult horses, but not in neonatal foals. In this study, we investigated extended-spectrum β-lactamase *Enterobacteriaceae* (ESBL-E) shedding and infection in hospitalized mares and their neonatal foals. Overall, we sampled rectal swabs from 55 pairs of mares and their foals on admission, and 33 of them were re-sampled on the 3rd day of hospitalization. We also collected clinical samples, when available. We found that shedding rates and bacterial species diversity increased significantly during hospitalization, both in mares and foals. On admission to hospital, foals' shedding was associated with umbilical infection. During hospitalization, it was associated with ampicillin treatment. Foals' shedding was independent of their mares' shedding. Four foals were infected with ESBL-E strains, including umbilical infections and wounds. We suggest further investigation and surveillance of ESBL-E in neonatal foals, in order to reduce resistance rates and infections.

Abstract: Extended-spectrum β-lactamase *Enterobacteriaceae* (ESBL-E) have been investigated in adult horses, but not in foals. We aimed to determine shedding and infection in neonatal foals and mares. Rectal swabs were sampled from mare and foal pairs on admission and on the 3rd day of hospitalization; enriched, plated, and bacteria were verified for ESBL production. Identification and antibiotic susceptibility profiles were determined (Vitek2). Genotyping was performed by multilocus sequence typing (MLST). Genes were identified by PCR and Sanger sequencing. Medical data were analyzed for risk factors (SPSS). On admission, 55 pairs were sampled, of which 33 pairs were re-sampled. Shedding rates on admission in foals and mares were 33% (95% CI 21–47%) and 16% (95% CI 8–29%), respectively, and during hospitalization, these increased significantly to 85% (95% CI 70–94%) and 58% (95% CI 40–73%), respectively. Foal shedding was associated with umbilical infection on admission ($P = 0.016$) and with ampicillin treatment during hospitalization ($p = 0.011$), and was independent of the mare's shedding. The most common ESBL-E was *Escherichia coli*. During hospitalization, species diversity increased. Four foals were infected with ESBL-E strains, including umbilical infections and wounds. This study substantiates an alarming prevalence of shedding in neonatal foals, which should be further investigated in order to reduce resistance rates.

Keywords: equine; foal; ESBL-E; antibiotic resistance; shedding; umbilical infection; risk factors

1. Introduction

Multidrug-resistant (MDR) *Enterobacteriaceae* are becoming a major worldwide concern in veterinary medicine, mainly due to the production of extended-spectrum β-lactamases (ESBLs) [1]. These widespread enzymes confer resistance to all the extended-spectrum cephalosporins and aztreonam, but not to cephamycins or carbapenems, and are usually inhibited by β-lactamase inhibitors [2]. ESBL genes are mainly plasmid-encoded and may co-carry additional antimicrobial resistances, including aminoglycosides, sulfa-derivatives, trimethoprim and quinolone resistance [3]. Therefore, treatment options of infections caused by ESBL-producing bacteria are limited. In human medicine, ESBL-E infection is associated with increased morbidity, mortality, length of hospital stay, delay of targeted appropriate treatment and higher costs [4,5]. Moreover, ESBL-E colonization has been identified as a risk factor for ESBL-E infection [6].

In horses, penicillin and cephalosporins are commonly prescribed [1], and antimicrobial resistance is of concern in a wide range of equine pathogens, including *Enterobacteriaceae* [7]. Reports of wounds, as well as respiratory and urinary tract infections, caused by ESBL-E are increasing in equine clinics [8,9]. Neonatal foals are considered as high-risk population due to their high-susceptibility and incomplete maturity of their immune system [10]. Although antibiotic resistance patterns in neonatal foals have been described [11], data on ESBL-E colonization and infections is still scarce. The motivation for this prospective study was the increasing prevalence of antibiotic resistant Gram-negative bacterial infections in hospitalized foals and the necessity to understand their origin. We hypothesized that foal ESBL-E shedding would be associated with mare shedding, as well as with specific clinical presentations and prior antibiotic treatment. We aimed to determine shedding prevalence among foal and mare pairs admitted to our hospital, to elucidate risk factors, and to determine clinical consequences.

2. Materials and Methods

2.1. Equine Study Population, Study Design and Sampling Methods

This prospective study was performed in the KSVM-VTH (Koret School of Veterinary Medicine—Veterinary Teaching Hospital). The study was approved by the Internal Research Review Committee of the KSVM-VTH (Reference number: KSVM-VTH/15_2015). Rectal swabs were collected from mares and their neonatal foals (pairs) on admission to the hospital over the course of one foaling season (November 2015–June 2016) [12]. Pairs of mares with foals under 30 days of age were included. Rectal sampling was performed immediately upon admission, prior to any medical treatment in the hospital. Rectal swabs were collected with owner consent. When both mare and foal survived and were not discharged, a second sample was taken on day 3 post admission. Overall, 55 pairs were sampled on admission, of which 33 (60%) pairs were re-sampled. When infection was detected clinically, as in umbilical infections for example, clinical samples were collected from the infection sites and *Enterobacteriaceae* isolates were tested for ESBL production. Sepsis was defined as a sepsis score greater than 11 [13], and umbilical infection was based on ultrasound and gross appearance [14].

2.2. Demographic and Medical Data

Medical records were reviewed for the following information: signalment (age, sex and breed) of mares and foals, parity, weight of foals, white blood cell count on admission, clinical signs on admission and during hospitalization, antibiotic therapy before and during hospitalization, surgical procedures, hospitalization length, short-term outcome and re-hospitalization.

2.3. ESBL-Producing Enterobacteriaceae (ESBL-E) Isolation and Species Identification

Rectal specimens were collected using bacteriological swabs (Meus s.r.l., Piove di Sacco, Italy) and were inoculated directly into a Luria Bertoni infusion enrichment broth (Hy-Labs, Rehovot, Israel) to increase sensitivity of ESBL-E detection [15]. After incubation at 37 °C (18–24 h), enriched samples were plated onto Chromagar ESBL plates (Hy-Labs, Rehovot, Israel), at 37 °C for 24 h. Pure isolates were stored at −80 °C for further analysis.

All isolates, from both rectal and clinical samples, were subjected to Vitek-2 (BioMérieux, Inc., Marcy-l'Etoile, France) for species identification and antibiotic susceptibility testing (AST-N270 Vitek 2 card). Chloramphenicol and doxycycline susceptibilities were analyzed using disc diffusion assay (Oxoid, Basingstoke, UK). ESBL-production was confirmed by combination disk diffusion using cefotaxime and ceftazidime discs (Oxoid, Basingstoke, UK), as well as cefotaxime and ceftazidime with clavulanic acid (Sensi-Discs BD, Breda, the Netherlands). Results were interpreted according to the Clinical and Laboratory Standards Institute (CLSI) guidelines [16]. Multidrug-resistant bacteria were defined as such due to their in vitro resistance to 3 or more classes of antimicrobial agents [17].

2.4. Molecular Characterization of ESBL-E

Isolates were examined for presence of the blaCTX-M group using a multiplex polymerase chain reaction (PCR) from ESBL-E DNA lysates, as previously described [18]. Isolates that were found to be blaCTX-M PCR negative were further examined for blaOXA-1, blaOXA2, blaOXA10 [19], blaTEM and blaSHV groups [20]. ESBL-producing *Escherichia coli* isolates were subjected to PCR in order to determine the presence of the pandemic *E. coli* ST131 strain [21]. All clinical isolates, as well as the additional fecal isolates from the same foals and their mares, were genotyped using an enterobacterial repetitive intergenic consensus (ERIC) PCR amplification using the following primer: 5′– AAGTAAGTGACTGGGGTGAGCG – 3′ [22]. Results were analyzed using GelJ software [23], and representative strains of each ERIC type were further analyzed by MLST for *E. coli*, *Klebsiella pneumonia* and *Klebsiella oxytoca*, as described before [24–26].

2.5. Statistical Analysis

Descriptive statistics were used to describe the shedding rates on admission and during hospitalization. Confidence intervals (95%) were calculated by Fisher's (WinPEPI 11.15 Describe A). Risk assessment was performed using Chi square or Fisher's exact tests for association between individual variables and shedding. Distribution of continuous parameters was evaluated by Shapiro-Wilk test and all were subsequently analyzed for statistical significance between two groups by Mann-Whitney U test due to non-parametric distribution. The agreement between mare and foal shedding status was analyzed by Cohen's kappa for agreement beyond chance using the on-line Vassarstat Kappa Calculator (http://vassarstats.net/kappa.html). The McNemar test was used to examine the significance of matched pairs (shedding on admission and during hospitalization). $p <$ 0.05 was considered statistically significant. A logistic regression model (multivariable analysis) was conducted using variables with $p < 0.10$, using the ENTER method (IBM SPSS Statistics 23).

3. Results

3.1. Characterization of Equine Study Population

The 55 mare–foal pairs included in this study represent 83% of the total number of mare–foal pairs admitted to the large animal department during the study period (other pairs were not sampled for logistical reasons). Out of nine pairs that were re-hospitalized, two were re-sampled on second admission, due to suspected infections (a dermal abscess in foal #1 and an umbilical infection in foal #2). Out of the sampled foals, n = 41 (74.5%) were fillies. Horses represented diverse breeds (Supplementary Information Table S1), including 60% Arabian horse (n = 66), 14.5% Tennessee walker (n = 16), 13.6% Quarter Horse (n = 15), 3.6% Single Footing (n = 4), 2.7% Appaloosa (n = 3), as well as one pair each of Missouri Fox Trotter, Friesian and Miniature breeds. Mares' median age was 6 years, and foals' median age was 3 days. Signalment was not consistent with the general hospital population, as these mares represent the brood mare population, along with their foals.

Admission to the hospital was due to disease of the foal, in most instances (n = 53/55, 96%). Two foals (n = 2/55, 4%) were healthy and were referred to the hospital due to disease of the mare. The pathologies were diverse (Table 1). One third of foals (n = 18/55, 33%) received antibiotics prior to hospital admission. As for the mares, 87% (n = 48/55) were healthy on arrival. One mare (n = 1/55, 1.8%) received antibiotics prior to hospitalization. For 43% of mares (n = 24/55), this was the first parturition.

Table 1. Pathologies of foals and mares on admission and during hospitalization.

Pathology	No. of Horses (%)	
	On Admission	Developed during Hospitalization
Foals	n = 55	n = 33
Diarrhea	13 (24)	4 (12)
Umbilical infection	13 (24)	0
Sepsis	12 (22)	0
Prematurity	10 (17)	0
Septic polyarthritis	9 (16)	0
Orthopedic problems (other than septic polyarthritis)	9 (15)	1 (3)
Perinatal Asphyxia Syndrome (PAS)	8 (13)	0
Respiratory problems	6 (11)	0
Colic	6 (10)	1 (3)
Injury	3 (5)	0
Neurological signs (other than PAS)	1 (2)	4 (12)
Uroperitoneum	1 (2)	1 (3)
Phlebitis	0	1 (3)
Uveitis	0	2 (6)
Peritonitis	0	1 (3)
Other (hernia, guttural pouch tympany and piroplasmosis)	3 (5)	0
Mares	n = 55	n = 33
Colic	2 (4)	2 (6)
Retained placenta	2 (4)	0
Injury	1 (2)	0
Orthopedic syndromes	1 (2)	0
Placentitis	1 (2)	0
Colitis	0	1 (3)

3.2. Hospital Procedures, Antibiotic Therapy and Outcome

During hospitalization, 95% of foals (n = 52/55) and 5% of mares (n = 3/55) were treated with antibiotics. Empirical antibiotic treatment of the foals included a combination of ampicillin and amikacin [11,27] for broad-spectrum coverage. Forty percent (n = 22/55) of foals underwent a surgical procedure and 25% (n = 14/55) had a urinary catheter inserted during hospitalization. Two thirds of foals (n = 36/55, 66%) were discharged, 27% (n = 15/55) died or were euthanized, and 7% (n = 4/55) were discharged contrary to medical advice. The median hospitalization duration was 3 days (range, 0–32 days). Nine foals (25%) that were discharged were re-admitted to the hospital within one month of discharge.

3.3. Prevalence of ESBL-E Shedding among Foals and Mares

Shedding rates of ESBL-E in foals and mares on admission were 33% (95% CI 21–47%) and 16% (95% CI 8–29%), respectively, and were not different ($p = 0.075$). Most of the shedding foals (n = 13/18) upon admission were accompanied by a non-shedding mare. In both populations, shedding rates increased significantly during hospitalization from 33% to 85% (95% CI 70–94%) in foals and from 16% to 58% (95% CI 40–73%) in mares (Table 2). The difference in shedding rates between mares and foals during hospitalization was significant ($p = 0.028$). Nineteen out of 22 non-shedding foals on admission (86%) were re-sampled and acquired ESBL-E during hospitalization. Ten of 18 shedding foals on admission (56%) remained hospitalized and were re-sampled. Nine of them (90%) remained positive and one turned negative. As for mares, five of nine that shed on admission remained hospitalized and remained positive during hospitalization. Fourteen mares, out of 28 negative mares on admission that were re-sampled, acquired ESBL-E (50%). Therefore, shedding rates increased significantly during hospitalization in both mares and foals ($p < 0.01$ for both populations).

Table 2. Shedding rates of ESBL-E in mares and foals on admission and during hospitalization.

	on Admission [1]			≥ 72 h of Hospitalization [2]		
Horses	**Shedding (%)**	**Total No. of ESBL-E Isolates**	***bla*ESBL Genes (%) [3]**	**(%)Shedding**	**Total No. of ESBL-E Isolates**	***bla*ESBL Genes (%) [3]**
Foals	18/55 (33) (95% CI 21–47)	18	*Bla*CTXM-1: 14/19 (74) *Bla*CTXM-9: 2/19 (11)	28/33 (85) [7] (95% CI 70–94)	46 [4]	CTX-M-1: 31/46 (67) CTX-M-2: 1/46 (2) OXA-1: 3/46 (7)
Mares	9/55 (16) (95% CI 8–29)	11 [5]	*Bla*CTXM-1: 6/11 (55)	19/33 (58) [8] (95% CI 40–73)	27 [6]	CTX-M-1: 16/27 (59) CTX-M-2: 1/27 (4) CTX-M-9: 2/27 (7) TEM-163: 2/27 (7)

[1] Rectal swabs were collected immediately on admission. [2] A second rectal swab was collected from all foals and mares that remained hospitalized. [3] ESBL genes were not identified in all ESBL-E. [4] Ten foals shed one ESBL-E; 15 shed two ESBL-E; two shed three ESBL-E isolates. [5] Eight mares shed one ESBL-E and one mare shed three ESBL-E isolates. [6] Eleven mares shed one ESBL-E; five shed two ESBL-E; two shed three ESBL-E isolates. [7] Foal shedding rates increased significantly following hospitalization and were significantly higher compared to mare shedding rates during hospitalization. [8] Mare shedding rates increased significantly following hospitalization.

3.4. Species Distribution of ESBL-E Shedding Isolates

Overall, 127 ESBL-E bacterial isolates were analyzed (Supplementary Information Table S2). The major fecal bacterial species on admission was *E. coli* (88% and 73% in foals and mares, respectively, Figure 1A,B). During hospitalization, the diversity of ESBL-E species increased in both populations (Figure 1C,D), with the following species distribution in foals and mares: *E. coli*—44% and 52%, respectively; *Klebsiella pneumoniae*—30% and 22%, respectively; and *Enterobacter cloacae*—13% and 11%, respectively. During hospitalization, ESBL-*Salmonella enterica* isolates were identified, consisting of 7% (3/46) and 7% (2/27) of ESBL-E shed by foals and mares, respectively. In addition, 25% (5/20) and 29% (4/14) of foals and mares, respectively, that shed ESBL-*E. coli* during hospitalization initially shed the same species on admission.

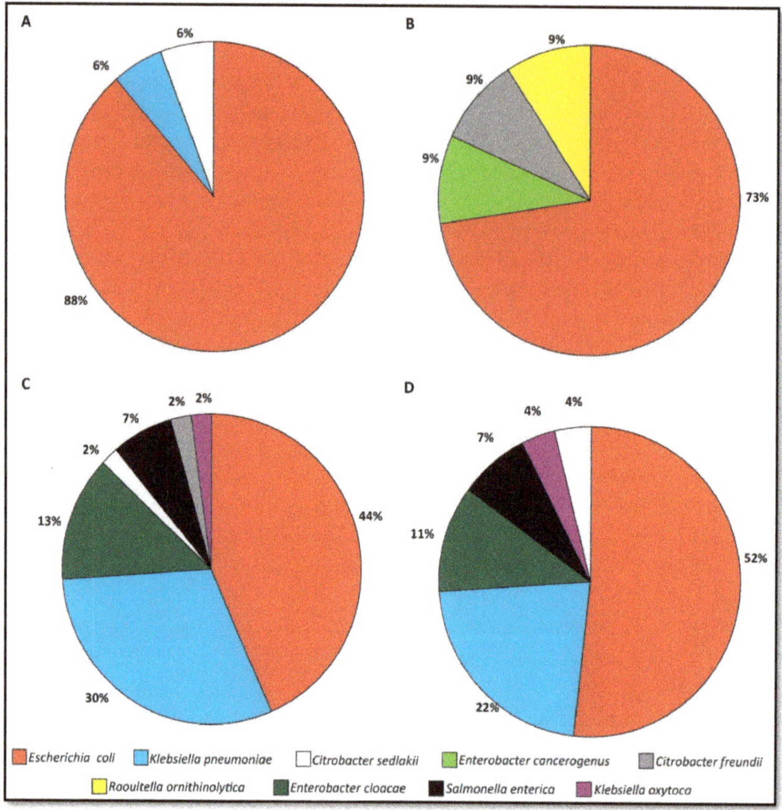

Figure 1. Species distribution of ESBL-E shedding on admission; foals (**A**, n = 18 isolates) and mares (**B**, n = 12 isolates); and 72 h post admission, foals (**C**, n = 46 isolates) and mares (**D**, n = 27 isolates).

The main *bla*ESBL gene group identified in ESBL-E isolates was CTX-M-1 (Table 2). Molecular screening for the presence of pandemic *E. coli* sequence type ST131 genetic lineage among all ESBL-*E. coli* isolates (n = 66) was negative.

3.5. Antibiotic Susceptibility Profiles of ESBL-E Fecal Isolates

3.5.1. ESBL-E Isolates on Admission

Antibiotic resistance rates within foal and mare populations varied, mainly with respect to ciprofloxacin, ofloxacin (27% and 25% in foals vs. 0% in mares), amikacin (0% in foals vs. 8% in mares) and gentamicin (33% in foals vs. 50% in mares), with significantly higher resistance rates to gentamicin compared to amikacin in foals ($p < 0.05$) (Figure 2A,B).

3.5.2. ESBL-E Isolates during Hospitalization

ESBL-E isolates recovered from hospitalized animals showed higher resistance rates compared to isolates on admission. Significant increase in resistance rates against amikacin and gentamicin were detected in strains isolated from foals ($p < 0.05$). All isolates were susceptible to carbapenems (Figure 2C,D).

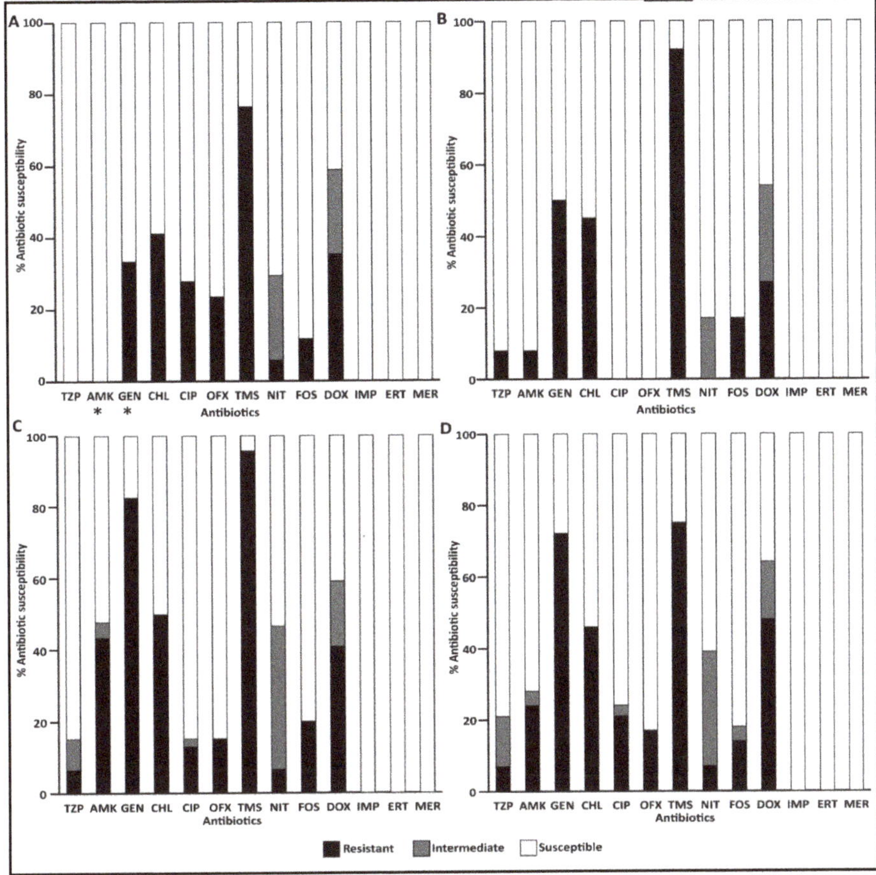

Figure 2. Antibiotic susceptibility profiles of ESBL-E shed on admission; foals (**A**, n = 18 isolates) and mares (**B**, n = 11 isolates); and 72 h post admission, foals (**C**, n = 46 isolates) and mares (**D**, n = 29 isolates). Significant changes are marked with asterisks.

3.5.3. ESBL-E Isolates from Clinical Samples of Infection Sites

A total of 24 clinical samples were collected from 18 foals (6 foals had two clinical samples). Clinical samples were obtained aseptically from 10 joints, 11 umbilical samples (during omphalectomy), two wounds and one tendon sheath. Four out of 18 foals with any clinical infection (22%) had an infection with an ESBL-E strain. Overall, eight ESBL-E clinical isolates were recovered (Table 3). Five of them (63%) were identified as *E. coli*, two as *K. pneumoniae* and one as *S. enterica* (Table 3). Along with ESBL-producing *E. coli*, an MDR *Acinetobacter baumannii* strain was recovered from a wound infection sampled from the leg of foal #2. All foals with clinical infection with ESBL-E shed at least one ESBL-E, either on admission or 72-hours post admission. In two of the foals (#1 and #3) the same ESBL-E strain that was isolated from the clinical sample was also found in the fecal sample (Table 3).

Table 3. ESBL-E from clinical and samples isolated from four foals during hospitalization.

Foal	Age on Admission	ESBL-E Shedding Status				Clinical ESBL-E Infection		
		1st Admission	1st Hospitalization	2nd Admission	2nd Hospitalization	ESBL-E Species	Source	Outcome
1	<12 h	Negative	*E. coli* ST88	*K. pneumoniae* ST1552	Not sampled	*K. pneumoniae* ST1552	abscess [1]	Discharged
2	<12 h	Negative	*E. coli* ST38	*E. coli* ST86 *K. oxytoca* ST194 *S. enterica*	*Enterobacter cloacae*	*E. coli* ST746 *E. coli* ST746	umbilicus [2] wound	Euthanized
3	<12 h	Negative	*E. coli* ST746 *K. pneumoniae* ST37	No second hospitalization		*E. coli* ST746 *K. pneumoniae* ST585 *S. enterica*	umbilicus [3]	Euthanized
4	17 d	*E. coli* [4]	Discharged	No second hospitalization		*E. coli* ST69 *E. coli* ST69	wound [5] umbilicus	Discharged

[1] Sampled on second admission, 27 days after first hospitalization. [2] Sampled on second admission, 6 days after first hospitalization, which prolonged 4 days. The wound developed 9 days after second admission. [3] Sampled following 8 hospitalization days. [4] The bacteria were not recovered for further analysis. [5] Sampled on admission. The foal suffered from infected umbilicus and wound for a week before admission.

All ESBL-E from clinical isolates in the study were MDR, as they also showed resistance to aminoglycosides and trimethoprim-sulpha, but were all susceptible to quinolones. In seven out of eight isolates, the ESBL gene was CTX-M group 1, and in one isolate, the ESBL gene was CTX-M group 9. Susceptibility profiles of individual isolates are displayed in Supplementary Information Table S1.

3.6. Risk Factor Analysis for ESBL-E Shedding

Cohen's Kappa for the agreement beyond chance between the ESBL-E shedding status of the foal and that of the mare was 0.2894 (95% CI 0–0.5961) on admission and was 0.2546 (95% CI 0–0.6143) during hospitalization. Both results are interpreted as minimal agreement [28]. On admission, a significant association was identified between foal's shedding and umbilical infection on admission ($p = 0.016$, 8 foals suffered from umbilical infection, out of 18 shedding foals). Odds ratio for umbilical infection in shedding foals on admission was 5.5 (95% CI 1.21–18.97). All other associations resulted in $p > 0.10$ (Supplementary Information Table S3); therefore, multivariable analysis for foal's shedding on admission was not conducted.

During hospitalization, foal's shedding was significantly associated with ampicillin treatment ($p = 0.002$, 26 foals were treated, out of 28 shedding foals). In a multivariate analysis for foals' shedding during hospitalization, the model included the following parameters ($p < 0.10$): hyperthermia on arrival, ampicillin treatment during hospitalization, diarrhea during hospitalization, and length of stay. The only significant risk factor was ampicillin treatment during hospitalization ($p = 0.011$, OR = 36.88, 95% CI 2.25–603.26).

4. Discussion

This is the first study to investigate shedding and infection with ESBL-E in hospitalized foals and to identify the risk factors involved. Shedding rates found on admission in foals and mares were high (33% and 16%, respectively) and increased significantly during hospitalization (85% and 58% in foals and mares, respectively). In addition to identifying an alarming prevalence and incidence, neonatal foals' ESBL-E shedding was associated with umbilical infection on admission and with ampicillin treatment during hospitalization.

Previous reports on ESBL-producing *E. coli* shedding/colonization rates in adult horses in the community ranged between 6.3 and 9% [8,29], and were lower than 10% on admission to a hospital [30]. Overall, data regarding shedding/colonization of ESBL-E in equine populations is limited and describes only ESBL-producing *E. coli*. Therefore, it is possible that the prevalence of all ESBL-E is higher, although in this study, *E. coli* was the main pathogen. ESBL-E is known to be endemic to Israel in human medicine [31], and therefore, a higher prevalence of ESBL-E circulating in the community setting may also explain the high rate on admission to the hospital. In a recent study, a high shedding/colonization rate (23.7%) was found in cattle in 40 farms in Israel. The mean prevalence of ESBL-E shedding/colonization was the highest in calves and gradually declined with maturation in adult cows [32]. In different countries, equine shedding/colonization rates may be different and generalizing our results should be done with caution.

The significant increase in shedding rates during hospitalization is alarming. Our findings support previous studies that describe a significant increase in prevalence during hospitalization [33], which was suggested to be associated with hospitalization length, mixed-purpose hospitalization yards [30] and high antimicrobial usage even in untreated animals [29]. This may explain the increase in mare ESBL-E shedding status, although they were mainly healthy. Alongside an increase in ESBL-E shedding rates, differences were observed in ESBL-E species distribution and in antibiotic resistance profiles of ESBL-E recovered from hospitalized foals and mares (Figures 1 and 2). This finding supports nosocomial acquisition of ESBL-E from the hospital environment, acquisition of ESBL-encoding mobile genetic elements from other bacteria present in the gastrointestinal microbiota or due to treatment with antimicrobial drugs, as was suggested previously [30,32].

In this study, *S. enterica* was found, comprising 7% of all ESBL-E isolates in hospitalized mares and foals, and was the causative pathogen of an umbilical infection in one foal. This pathogen is highly concerning, mainly due to its zoonotic and outbreak potential. In the KSVM-VTH, all horses that are admitted with acute diarrhea are routinely isolated, and every horse that develops acute diarrhea during hospitalization is moved to the isolation ward immediately. In previous reports, different serotypes of the *Salmonella* genus were isolated from foals and identified as a cause of diarrhea and septic arthritis [34,35]. In addition, foals with gastrointestinal tract disease were 3.27 times as likely to be shedding *Salmonella* organisms compared to adult horses [36]. However, reports on ESBL-producing *Salmonella* species in neonatal foals is lacking, posing the need for further studies and the necessity of surveillance actions, including the implementation of extensive infection control measures in order to identify hot spots for acquisition.

As opposed to human newborns [37] and piglets [38], shedding of ESBL-E among foals upon hospital admission in our study was weakly associated with the ESBL-E shedding status of their mares. This result is intriguing because mares would be expected to serve as a direct source for infection of foals either in utero or during parturition, as reported in the case of women and neonates [39]. This does not seem to be the case in these mares and foals, although the wide confidence interval due to the small sample size must be noted. This finding could be related to, and affected by, differences in immunity between neonatal foals and mares and due to their high exposure to the environment in the stable. The findings of this study, supported by a previous report regarding ESBL-producing *E. coli* isolates from an equine stable [8], suggest that the equine environment is a source for ESBL-E acquisition in the foals rather than the mare. In contrast to the weak association between mare and foal shedding status, we found identical ESBL-E fecal strains mares and foals and infected foals, supporting transmission of ESBL-E strains between mares and their foals. To further understand the epidemiology of mare and foal transmission, longitudinal studies, including larger sample sizes, need to be performed.

Although the origin of bacterial infection in foal medicine is often undefined, it may be a leading cause of sepsis and death. ESBL-E may disseminate systemically, presumably through the intestinal tract of the foal as a route of invasion. Although not directly proven, bacteria may hypothetically cross the intestinal barrier into the interstitium, lymphatics and bloodstream. Reports documenting that Gram-negative enteric bacteria are the predominant isolates from neonatal foals with sepsis [34] provide further evidence of the importance of the gastrointestinal tract as a major bacterial portal of entry. Since all foals with ESBL-E-associated infections also shed ESBL-E, often with an identical strain, the association between shedding and infection is most likely. This study was prospectively designed to investigate one foaling season, in which we found four foals infected with ESBL-E. To better understand the connection between shedding and infection, longitudinal studies encompassing larger population are required.

Another clinical implication of our study concerns antibiotic treatment during hospitalization. In hospitalized foals, we identified treatment with ampicillin as significantly associated with ESBL-E shedding. This may explain the significant difference in shedding rate between foals and mares during hospitalization, as only foals were treated with ampicillin. However, there may be numerous reasons for the difference, such as maturity of the immune system. In human neonatal intensive care units, prior antibiotic treatment, combination of ampicillin/gentamicin and cephalosporin treatment were detected as risk factors for shedding and/or infection with ESBL-E [40]. In our study, most foals were treated with combination of ampicillin/amikacin [27] with or without other antibiotics; therefore, it was impossible to distinguish the effect of ampicillin as mono-therapy. In addition, treatment with cephalosporins was relatively rare and may be the reason no significant association was found. The association with ampicillin treatment should be further studied.

The limitations of this study include small sample size and retrospective medical data collection. The sample size w41as limited due to the number of admissions in the relevant foaling season. Even though the statistical analysis did reveal significant associations, a larger sample size may have resulted in additional associations.

This study underscores the importance of applying an active surveillance policy for ESBL-E shedding in foals. As we revealed the importance of ESBL-E diagnosis, future studies should include a larger cohort and further understanding of the source of these ESBL-producing strains, both on admission and in the hospital setting.

5. Conclusions

The results from this study substantiate the alarming occurrence of ESBL-E in equine neonatal medicine. Our data confirm that mares and their neonatal foals may shed and be infected by ESBL-E. Further studies and active surveillance should focus on community-onset, nosocomial ESBL-E shedding, and infection in foals, describing molecular characteristics and pathogenicity of ESBL-E.

Supplementary Materials: The following are available online at http://www.mdpi.com/2076-2615/9/9/600/s1, Table S1: Characterization of equine study population. Table S2: Antimicrobial susceptibility profiles of individual isolates. Susceptible = 0, intermediate susceptibility = 1, resistant = 2. Empty cells mean lack of susceptibility test results due to technical reasons. Table S3: Results of univariable analysis of variables gleaned from the medical records and evaluated for association with the outcome of ESBL-E shedding status of the individual animal.

Author Contributions: Conceptualization: A.S., S.N.-V. and A.S.-T.; Methodology: Y.P., H.A., K.K., A.S.-T.; Software: G.A.-S., and S.T.-L.; Validation: A.S., S.N.-V. and A.S.-T.; Formal Analysis: A.S., S.N.-V., A.S.-T., and G.A.-S.; Investigation: A.S.-T.; Resources: A.S., S.N.-V. and Y.P.; Data Curation: A.S.-T.; Writing – Original Draft Preparation: A.S.-T., A.S., S.N.-V.; Writing – Review & Editing: all authors; Visualization: A.S., S.N.-V. and A.S.-T.; Supervision, project administration and funding Acquisition: A.S. and S.N.-V.

Acknowledgments: Erez Hanael, for graphical assistance. The staff in the large animal department in the KSVM-VTH for sampling the animals.

Conflicts of Interest: The authors declare no conflict of interest.

References

1. Rubin, J.E.; Pitout, J.D.D. Extended-spectrum β-lactamase, carbapenemase and AmpC producing Enterobacteriaceae in companion animals. *Vet. Microbiol.* **2014**, *170*, 10–18. [CrossRef] [PubMed]

2. Lee, J.H.; Bae, I.K.; Lee, S.H. New definitions of extended-spectrum β-lactamase conferring worldwide emerging antibiotic resistance. *Med. Res. Rev.* **2012**, *32*, 216–232. [CrossRef] [PubMed]

3. Liu, X.; Thungrat, K.; Boothe, D.M. Occurrence of OXA-48 Carbapenemase and Other β-Lactamase Genes in ESBL-Producing Multidrug Resistant Escherichia coli from Dogs and Cats in the United States, 2009–2013. *Front. Microbiol.* **2016**, *7*. [CrossRef] [PubMed]

4. Denkel, L.A.; Gastmeier, P.; Piening, B. To screen or not to screen mothers of preterm infants for extended-spectrum beta-lactamase-producing Enterobacteriaceae (ESBL-E). *J. Perinatol.* **2015**, *35*, 893–894. [CrossRef] [PubMed]

5. Schwaber, M.J.; Navon-Venezia, S.; Kaye, K.S.; Ben-Ami, R.; Schwartz, D.; Carmeli, Y. Clinical and economic impact of bacteremia with extended- spectrum-beta-lactamase-producing Enterobacteriaceae. *Antimicrob. Agents Chemother.* **2006**, *50*, 1257–1262. [CrossRef] [PubMed]

6. Bert, F.; Larroque, B.; Paugam-Burtz, C.; Dondero, F.; Durand, F.; Marcon, E.; Belghiti, J.; Moreau, R.; Nicolas-Chanoine, M.-H. Pretransplant Fecal Carriage of Extended-Spectrum β-Lactamase–producing Enterobacteriaceae and Infection after Liver Transplant, France. *Emerg. Infect. Dis.* **2012**, *18*, 908–916. [CrossRef]

7. Weese, J.S. Antimicrobial use and antimicrobial resistance in horses. *Equine Vet. J.* **2015**, *47*, 747–749. [CrossRef]

8. Dolejska, M.; Duskova, E.; Rybarikova, J.; Janoszowska, D.; Roubalova, E.; Dibdakova, K.; Maceckova, G.; Kohoutova, L.; Literak, I.; Smola, J.; et al. Plasmids carrying blaCTX-M-1 and qnr genes in Escherichia coli isolates from an equine clinic and a horseback riding centre. *J. Antimicrob. Chemother.* **2011**, *66*, 757–764. [CrossRef]

9. Walther, B.; Klein, K.-S.; Barton, A.-K.; Semmler, T.; Huber, C.; Wolf, S.A.; Tedin, K.; Merle, R.; Mitrach, F.; Guenther, S.; et al. Extended-spectrum beta-lactamase (ESBL)-producing Escherichia coli and Acinetobacter baumannii among horses entering a veterinary teaching hospital: The contemporary "Trojan Horse". *PLoS ONE* **2018**, *13*, e0191873. [CrossRef]

10. Wohlfender, F.D.; Barrelet, F.E.; Doherr, M.G.; Straub, R.; Meier, H.P. Diseases in neonatal foals. Part 1: the 30 day incidence of disease and the effect of prophylactic antimicrobial drug treatment during the first three days post partum. *Equine Vet. J.* **2009**, *41*, 179–185. [CrossRef]

11. Theelen, M.J.P.; Wilson, W.D.; Edman, J.M.; Magdesian, K.G.; Kass, P.H. Temporal trends in prevalence of bacteria isolated from foals with sepsis: 1979-2010. *Equine Vet. J.* **2014**, *46*, 169–173. [CrossRef] [PubMed]

12. Dyakova, E.; Bisnauthsing, K.N.; Querol-Rubiera, A.; Patel, A.; Ahanonu, C.; Tosas Auguet, O.; Edgeworth, J.D.; Goldenberg, S.D.; Otter, J.A. Efficacy and acceptability of rectal and perineal sampling for identifying gastrointestinal colonization with extended spectrum β-lactamase Enterobacteriaceae. *Clin. Microbiol. Infect.* **2017**, *23*, 577.e1–577.e3. [CrossRef] [PubMed]

13. Brewer, B.D.; Koterba, A.M.; Carter, R.L.; Rowe, E.D. Comparison of empirically developed sepsis score with a computer generated and weighted scoring system for the identification of sepsis in the equine neonate. *Equine Vet. J.* **1988**, *20*, 23–24. [CrossRef] [PubMed]

14. Oreff, G.L.; Tatz, A.J.; Dahan, R.; Segev, G.; Berlin, D.; Kelmer, G. Surgical management and long-term outcome of umbilical infection in 65 foals (2010-2015). *Vet. Surg.* **2017**, *46*, 962–970. [CrossRef] [PubMed]

15. Murk, J.-L.A.N.; Heddema, E.R.; Hess, D.L.J.; Bogaards, J.A.; Vandenbroucke-Grauls, C.M.J.E.; Debets-Ossenkopp, Y.J. Enrichment broth improved detection of extended-spectrum-beta-lactamase-producing bacteria in throat and rectal surveillance cultures of samples from patients in intensive care units. *J. Clin. Microbiol.* **2009**, *47*, 1885–1887. [CrossRef]

16. Clinical and Laboratory Standards Institute (CLSI). *Performance Standards for Antimicrobial Susceptibility Testing*, 26th ed.; Clinical and Laboratory Standards Institute: Wayne, PA, USA, 2016.

17. Falagas, M.E.; Karageorgopoulos, D.E. Pandrug Resistance (PDR), Extensive Drug Resistance (XDR), and Multidrug Resistance (MDR) among Gram-Negative Bacilli: Need for International Harmonization in Terminology. *Clin. Infect. Dis.* **2008**, *46*, 1121–1122. [CrossRef]

18. Woodford, N.; Fagan, E.J.; Ellington, M.J. Multiplex PCR for rapid detection of genes encoding CTX-M extended-spectrum β-lactamases. *J. Antimicrob. Chemother.* **2006**, *57*, 154–155. [CrossRef]

19. Lin, S.-P.; Liu, M.-F.; Lin, C.-F.; Shi, Z.-Y. Phenotypic detection and polymerase chain reaction screening of extended-spectrum β-lactamases produced by Pseudomonas aeruginosa isolates. *J. Microbiol. Immunol. Infect.* **2012**, *45*, 200–207. [CrossRef]

20. Tofteland, S.; Haldorsen, B.; Dahl, K.H.; Simonsen, G.S.; Steinbakk, M.; Walsh, T.R.; Sundsfjord, A.; Norwegian ESBL Study Group. Effects of phenotype and genotype on methods for detection of extended-spectrum-beta-lactamase-producing clinical isolates of Escherichia coli and Klebsiella pneumoniae in Norway. *J. Clin. Microbiol.* **2007**, *45*, 199–205. [CrossRef]

21. Johnson, J.R.; Clermont, O.; Johnston, B.; Clabots, C.; Tchesnokova, V.; Sokurenko, E.; Junka, A.F.; Maczynska, B.; Denamur, E. Rapid and Specific Detection, Molecular Epidemiology, and Experimental Virulence of the O16 Subgroup within Escherichia coli Sequence Type 131. *J. Clin. Microbiol.* **2014**, *52*, 1358–1365. [CrossRef]

22. Versalovic, J.; Koeuth, T.; Lupski, J.R. Distribution of repetitive DNA sequences in eubacteria and application to fingerprinting of bacterial genomes. *Nucleic Acids Res.* **1991**, *19*, 6823–6831. [CrossRef] [PubMed]

23. Heras, J.; Domínguez, C.; Mata, E.; Pascual, V.; Lozano, C.; Torres, C.; Zarazaga, M. GelJ – a tool for analyzing DNA fingerprint gel images. *BMC Bioinforma.* **2015**, *16*, 270. [CrossRef] [PubMed]

24. Herzog, K.A.T.; Schneditz, G.; Leitner, E.; Feierl, G.; Hoffmann, K.M.; Zollner-Schwetz, I.; Krause, R.; Gorkiewicz, G.; Zechner, E.L.; Högenauer, C. Genotypes of Klebsiella oxytoca isolates from patients with nosocomial pneumonia are distinct from those of isolates from patients with antibiotic-associated hemorrhagic colitis. *J. Clin. Microbiol.* **2014**, *52*, 1607–1616. [CrossRef] [PubMed]

25. Wirth, T.; Falush, D.; Lan, R.; Colles, F.; Mensa, P.; Wieler, L.H.; Karch, H.; Reeves, P.R.; Maiden, M.C.J.; Ochman, H.; et al. Sex and virulence in Escherichia coli: an evolutionary perspective. *Mol. Microbiol.* **2006**, *60*, 1136–1151. [CrossRef] [PubMed]

26. Diancourt, L.; Passet, V.; Verhoef, J.; Grimont, P.A.D.; Brisse, S. Multilocus Sequence Typing of Klebsiella pneumoniae Nosocomial Isolates. *J. Clin. Microbiol.* **2005**, *43*, 4178–4182. [CrossRef]

27. Palmer, J. Update on the management of neonatal sepsis in horses. *Vet. Clin. North Am. Equine Pract.* **2014**, *30*, 317–336. [CrossRef]

28. McHugh, M.L. Interrater reliability: the kappa statistic. *Biochem. Med.* **2012**, *22*, 276–282. [CrossRef]

29. Maddox, T.W.; Clegg, P.D.; Diggle, P.J.; Wedley, A.L.; Dawson, S.; Pinchbeck, G.L.; Williams, N.J. Cross-sectional study of antimicrobial-resistant bacteria in horses. Part 1: Prevalence of antimicrobial-resistant Escherichia coli and methicillin-resistant Staphylococcus aureus. *Equine Vet. J.* **2012**, *44*, 289–296. [CrossRef]

30. Maddox, T.W.; Williams, N.J.; Clegg, P.D.; O'Donnell, A.J.; Dawson, S.; Pinchbeck, G.L. Longitudinal study of antimicrobial-resistant commensal Escherichia coli in the faeces of horses in an equine hospital. *Prev. Vet. Med.* **2011**, *100*, 134–145. [CrossRef]

31. Paterson, D.L.; Bonomo, R.A. Extended-Spectrum β-Lactamases: a Clinical Update. *Clin. Microbiol. Rev.* **2005**, *18*, 657–686. [CrossRef]

32. Adler, A.; Sturlesi, N.A.; Fallach, N.; Zilberman-Barzilai, D.; Hussein, O.; Blum, S.E.; Klement, E.; Schwaber, M.J.; Carmeli, Y. Prevalence, Risk Factors, and Transmission Dynamics of Extended-Spectrum-β-Lactamase-Producing Enterobacteriaceae: a National Survey of Cattle Farms in Israel in 2013. *J. Clin. Microbiol.* **2015**, *53*, 3515–3521. [PubMed]

33. Damborg, P.; Marskar, P.; Baptiste, K.E.; Guardabassi, L. Faecal shedding of CTX-M-producing Escherichia coli in horses receiving broad-spectrum antimicrobial prophylaxis after hospital admission. *Vet. Microbiol.* **2012**, *154*, 298–304. [CrossRef] [PubMed]

34. Olivo, G.; Lucas, T.M.; Borges, A.S.; Silva, R.O.S.; Lobato, F.C.F.; Siqueira, A.K.; da Silva Leite, D.; Brandão, P.E.; Gregori, F.; de Oliveira-Filho, J.P.; et al. Enteric Pathogens and Coinfections in Foals with and without Diarrhea. *BioMed Res. Int.* **2016**, *2016*, 1512690. [CrossRef] [PubMed]

35. Barceló Oliver, F.; Russell, T.M.; Uprichard, K.L.; Neil, K.M.; Pollock, P.J. Treatment of septic arthritis of the coxofemoral joint in 12 foals. *Vet. Surg.* **2017**, *46*, 530–538. [CrossRef] [PubMed]

36. Ernst, N.S.; Hernandez, J.A.; MacKay, R.J.; Brown, M.P.; Gaskin, J.M.; Nguyen, A.D.; Giguere, S.; Colahan, P.T.; Troedsson, M.R.; Haines, G.R.; et al. Risk factors associated with fecal Salmonella shedding among hospitalized horses with signs of gastrointestinal tract disease. *J. Am. Vet. Med. Assoc.* **2004**, *225*, 275–281. [CrossRef] [PubMed]

37. Denkel, L.A.; Schwab, F.; Kola, A.; Leistner, R.; Garten, L.; von Weizsäcker, K.; Geffers, C.; Gastmeier, P.; Piening, B. The mother as most important risk factor for colonization of very low birth weight (VLBW) infants with extended-spectrum β-lactamase-producing Enterobacteriaceae (ESBL-E). *J. Antimicrob. Chemother.* **2014**, *69*, 2230–2237. [CrossRef] [PubMed]

38. Callens, B.; Faes, C.; Maes, D.; Catry, B.; Boyen, F.; Francoys, D.; de Jong, E.; Haesebrouck, F.; Dewulf, J. Presence of antimicrobial resistance and antimicrobial use in sows are risk factors for antimicrobial resistance in their offspring. *Microb. Drug Resist.* **2015**, *21*, 50–58. [CrossRef] [PubMed]

39. Simonsen, K.A.; Anderson-Berry, A.L.; Delair, S.F.; Davies, H.D. Early-Onset Neonatal Sepsis. *Clin. Microbiol. Rev.* **2014**, *27*, 21–47. [CrossRef]

40. Li, X.; Xu, X.; Yang, X.; Luo, M.; Liu, P.; Su, K.; Qing, Y.; Chen, S.; Qiu, J.; Li, Y. Risk factors for infection and/or colonisation with extended-spectrum β-lactamase-producing bacteria in the neonatal intensive care unit: a meta-analysis. *Int. J. Antimicrob. Agents* **2017**, *50*, 622–628. [CrossRef]

MDPI
St. Alban-Anlage 66
4052 Basel
Switzerland
Tel. +41 61 683 77 34
Fax +41 61 302 89 18
www.mdpi.com

Animals Editorial Office
E-mail: animals@mdpi.com
www.mdpi.com/journal/animals